T0320292

LEAN MANUFACTURING

Business Bottom-Line Based

LEAN MANUFACTURING

Business Bottom-Line Based

John X. Wang

CRC Press
Taylor & Francis Group
Boca Raton London New York

CRC Press is an imprint of the
Taylor & Francis Group, an **informa** business

CRC Press
Taylor & Francis Group
6000 Broken Sound Parkway NW, Suite 300
Boca Raton, FL 33487-2742

First issued in paperback 2019

© 2011 by Taylor & Francis Group, LLC
CRC Press is an imprint of Taylor & Francis Group, an Informa business

No claim to original U.S. Government works

ISBN-13: 978-1-4200-8602-7 (hbk)
ISBN-13: 978-0-367-38371-8 (pbk)

This book contains information obtained from authentic and highly regarded sources. Reasonable efforts have been made to publish reliable data and information, but the author and publisher cannot assume responsibility for the validity of all materials or the consequences of their use. The authors and publishers have attempted to trace the copyright holders of all material reproduced in this publication and apologize to copyright holders if permission to publish in this form has not been obtained. If any copyright material has not been acknowledged please write and let us know so we may rectify in any future reprint.

Except as permitted under U.S. Copyright Law, no part of this book may be reprinted, reproduced, transmitted, or utilized in any form by any electronic, mechanical, or other means, now known or hereafter invented, including photocopying, microfilming, and recording, or in any information storage or retrieval system, without written permission from the publishers.

For permission to photocopy or use material electronically from this work, please access www.copyright.com (http://www.copyright.com/) or contact the Copyright Clearance Center, Inc. (CCC), 222 Rosewood Drive, Danvers, MA 01923, 978-750-8400. CCC is a not-for-profit organization that provides licenses and registration for a variety of users. For organizations that have been granted a photocopy license by the CCC, a separate system of payment has been arranged.

Trademark Notice: Product or corporate names may be trademarks or registered trademarks, and are used only for identification and explanation without intent to infringe.

Library of Congress Cataloging-in-Publication Data

Wang, John X., 1962-
    Lean manufacturing : business bottom-lined based / John X. Wang.
        p. cm.
    Includes bibliographical references and index.
    ISBN 978-1-4200-8602-7
    1. Lean manufacturing. 2. Cost control. I. Title.

TS155.W3893 2011
658.5--dc22   2010005440

Visit the Taylor & Francis Web site at
http://www.taylorandfrancis.com

and the CRC Press Web site at
http://www.crcpress.com

To

Zurich, Switzerland

Vienna, Austria

College Park, Maryland

Erie, Pennsylvania

Ann Arbor, Michigan

Chicago, Illinois

Cedar Rapids, Iowa

Tucson, Arizona

The cities have engineered my dreams.

Contents

Preface

About 50 miles east of the Mississippi River, I found myself back in Galesburg, a western Illinois city of 34,000 and the birthplace of poet Carl Sandburg. In September 2004, Maytag closed Galesburg's refrigeration products factory, whose history goes back to the period when the farming industry was developing in the Midwest around 1904. Having taught Design for Lean Six Sigma courses for Maytag in Galesburg, I founded Carl Sandburg College's Center for Manufacturing Excellence, representing the beautiful city's continued pursuit for world-class manufacturing.

Today's manufacturing companies face unprecedented challenges in reliability, efficiency, flexibility, and cost saving because of:

- Escalating material and labor costs that trim margins
- Global competition that is driving down prices and lead times
- Ever increasing shareholder expectations for profit and sales growth.

Having transformed product design and manufacturing at companies ranging from Maytag and Visteon to General Electric, I see the limitation of traditional Lean manufacturing, which comes in the form of techniques regarding a particular tool or toolbox, yet the factory floor, like everything in the global community, is profoundly driven by business bottom lines.

Business executives, manufacturing leaders, and engineers often ask me how to revitalize Lean Six Sigma. Delivering real bottom-line results from manufacturing improvements has proven to be much harder than expected for most companies. Total quality management (TQM), zero-defect manufacturing, and business process reengineering have dropped off the landscape for taking much too long and failing to deliver the promised results. Lean Six Sigma is now experiencing the same fundamental difficulty under the waves of economic shocks and global competition. The following eight key features make this new book unique and timely:

1. Shifting the paradigm to a Lean manufacturing based on business bottom line, dramatic gains can be achieved systematically using focused improvement strategy and methodology presented by the book.

2. Based on business bottom line, transformed Lean manufacturing approaches improvement from the global, system perspective, rather than through trying to improve individual departments or functions in isolation.

3. The book will show you how to turn the difficult challenge of understanding leverage points for generating bottom-line results into a systematic process that can be used over and over.

4. Based on business bottom line, Lean manufacturing enables value flow through the factory most effectively by reallocating the most overloaded resources that determine the maximum flow rate of production and making value flow through these bottlenecks.

5. Facilitating manufacturers to optimize production through their critical bottleneck in order to meet market demand rapidly, the book outlines the case studies between world events and the manufacturing efficiency, and present Lean manufacturing strategies and techniques designed to accelerate responses to current and future events on the floors of the world's manufacturing facilities.

6. Based on the business bottom line, Lean manufacturing increases manufacturing income and reliability, and improves customer satisfaction and delivery performance consistently.

7. Streamlining with business bottom lines, the transformed Lean manufacturing offers a way forward for manufacturing and project operations coping with global competition and other market forces. The book will show you how to harness these market forces by following the effective rules to reduce production and inventory costs.

8. The book will show you how to revitalize Lean Six Sigma by aligning it with business bottom line and thus delivering results that your executives, business leaders, and customers are expecting.

Written in the Christmas holiday season, this is also a preface to a New Year. When manufacturing businesses boom, the influx of new payroll money creates jobs throughout the local economy, as company employees begin buying new homes, cars, and other goods and services. Because of this "multiplier effect," good Lean manufacturing also impacts the bottom line of every home. Like I am sharing the book with my wife and two sons, I expect you will share the book with your family, colleagues, and friends.

About the Author

Dr. John X. Wang, the founder and Chief Master Black Belt of the Lean Six Sigma Institute of Technology, directs Lean manufacturing to improve business bottom line. As a leading expert in reliability engineering, Six Sigma, and Lean manufacturing, Dr. Wang has taught training courses at Panduit Corp., Maytag Corp., Visteon Corp., and General Electric—Gannon University Co-Op Graduate Programs. Wang has served as Reliability Engineering and Design for Six Sigma manager at Maytag (where he led reliability engineering best practices and Design for Lean Six Sigma training), as a Six Sigma Master Black Belt certified by Visteon (where he led Design for Six Sigma training programs), and as a Six Sigma Black Belt certified by General Electric (where he led Design for Six Sigma best practice projects). Wang has authored and coauthored four engineering books and numerous professional publications on business communication, decision making under uncertainty, risk engineering and management, Six Sigma, reliability engineering, and systems engineering. Having taught engineering and professional courses at Gannon University and National Technological University, Wang has been speaking and presenting at various international and national engineering conferences, symposiums, professional meetings, seminars, and workshops. Wang is an American Society for Quality Certified Reliability Engineer and an International Quality Federation Certified Master Black Belt. He received a B.A. (1985) and M.S. from Tsinghua University, Beijing, China; and a Ph.D. (1995) from the University of Maryland at College Park. Wang lives in Vail, Arizona, with his wife and two sons.

1

Introduction: Five Stages of Lean Manufacturing

This chapter enables you to

- Understand the concept of Lean manufacturing in the context of industrial dynamics and the bullwhip effect
- Recognize that, facing the challenges of global climate and changes in the economy, risk engineering and management is the primary target of today's manufacturing businesses
- Know the three toolboxes and five stages for implementing Lean manufacturing

1.1 Lean Manufacturing

Lean manufacturing is the production of goods using less of everything compared to mass production: less waste, less human effort, less manufacturing space, less investment in tools, and less engineering time to develop a new product. Lean manufacturing is a generic process management philosophy derived mostly from the Toyota Production System (TPS) as well as other industrial best practices. Lean manufacturing is renowned for its focus on reduction of Toyota's original "seven wastes" in order to improve overall customer satisfaction. According to TPS, waste in a process is any activity that does not result in moving the process closer to the final output or adding value to the final output. The seven wastes are:

1. Overproduction—Overproduction is to manufacture an item before it is actually required. Overproduction is highly costly to a manufacturing plant because it prohibits the smooth flow of materials and actually degrades quality and productivity.
2. Excess inventory—Excess inventory tends to hide problems on the plant floor, which must be identified and resolved to improve operating performance. Excess inventory increases lead times, consumes productive floor space, delays the identification of problems, and inhibits communication.

3. Waiting—Whenever goods are not moving or being processed, the waste of waiting occurs. Much of a product's lead time is tied up in waiting for the next operation. Waiting is usually caused by poor design material flow and information flow.

4. Transportation—Transporting product between processes is a cost incursion that adds no value to the product. Excessive movement and handling cause damage and are an opportunity for quality to deteriorate.

5. Unnecessary motion—As compared to transporting materials, motion refers to the producer, worker, or equipment's movement, which could cause damage, fatigue, wear, and safety issues.

6. Overprocessing—Using more expensive resources than are needed for the task or adding design features that are not needed by customers. Expensive resources also encourage overproduction in order to recover the high cost of this equipment.

7. Defects—Quality defects impact to the business bottom line, resulting in rework or scrap and associated costs such as quarantining inventory, reinspecting, rescheduling, capacity loss, and so forth.

Besides its focus on the production process, Lean manufacturing is also viewed as a management technology for manufacturing cost-reduction, which includes:

- Cost reduction by design—Product development determines 80 percent of manufacturing cost. The concept/architecture phase alone determines 60 percent of manufacturing cost. Significant cost reductions can be achieved by design for manufacturability (DFM), design for testability, design for reliability, and design for Six Sigma.

- Production cost reduction—Lean manufacturing could double labor productivity, cutting production throughput times by 90 percent, reducing inventories by 90 percent, cutting errors and scrap in half. Significant cost reductions can be achieved by optimizing material flow, information flow, manufacturing process capability, and manufacturing process control.

- Overhead cost reduction—Standard products can be built to order without inventory and specials can be mass customized on demand. Significant cost reduction can be achieved by implementing value stream mapping and forecasting.

- Standardization cost reduction—Standard part lists can be fifty times less than proliferated lists due to economies of scale resulting from larger purchases. Standard parts are easier to get and fewer types need to be purchased. Significant cost reduction can be achieved by implementing part standardization and commonality between product platforms.

- Product line rationalization cost reduction—Products with low profit margins may be losing money on a total cost basis. Significant cost reduction can be achieved by implementing product line rationalization to eliminate or outsource low-profit products that have high overhead demands and cost.

- Supply chain management cost reduction—Supply chain management impacts inventory cost significantly. Significant cost reduction can be achieved by optimizing supply chain design and process control.

- Cost of poor quality reduction—The cost of poor quality (COPQ) can be 15 to 40 percent of revenue. COPQ can be greatly reduced by design for manufacturing reliability including highly accelerated stress screening (HASS), effective burn-in, and design of experiments for manufacturing.

Lean manufacturing is often linked with Six Sigma, which emphasizes reduction of process variation (or its converse smoothness). Lean Six Sigma is a business improvement methodology that combines tools from both Lean manufacturing and Six Sigma. Lean manufacturing focuses on manufacturing cost reduction, whereas traditional Six Sigma focuses on defect reduction. Combining the two improves manufacturing productivity and quality to deal with challenges such as the bullwhip effect described in the next section.

1.2 Industrial Dynamics and Bullwhip Effect

The *bullwhip effect* (or whiplash effect) is an observed phenomenon in forecast-driven supply chains including manufacturing processes. The concept has its roots in J. Forrester's industrial dynamics (1961). Because customer demand is rarely perfectly stable, businesses must forecast demand in order to properly plan production, inventory, and other resources. Forecasts are based on statistics, and they are rarely perfectly accurate. Because forecast errors are a given, companies often carry an inventory buffer called *safety stock*. Moving up the supply chain from end consumer, to manufacturing factory, and then to raw materials supplier, each supply chain participant has greater observed variation in demand and thus greater need for safety stock. In periods of rising demand, downstream participants will increase their orders. In periods of falling demand, orders will fall or stop in order to reduce inventory. As shown in Figure 1.1 and Figure 1.2, the effect is that variations are amplified as one moves upstream in the supply chain (further from the customer). Supply chain experts have recognized that the bullwhip effect is a problem

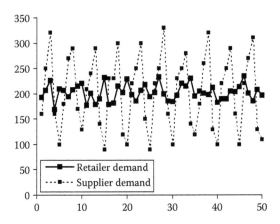

FIGURE 1.1
Bullwhip effect: Uncertainty amplification upstream of a supply (or downstream of a demand) chain.

in forecast-driven supply chains, and careful management of the effect is an important goal for production managers and supply chain managers.

The alternative is to establish a demand-driven supply chain that reacts to actual customer orders. In manufacturing, this concept is called Kanban, a method of communication within a just-in-time (JIT) system. Essentially, workers would use Kanban cards to communicate what have been completed and what the next operation will be in the process. At Dell, for example, the process is called *pull to order*. After getting an order, Dell notifies its suppliers about which components are needed, and the components are delivered within an hour and a half. Once the parts are delivered, the assembly-line process can begin prepping components. Dell then begins manufacturing the actual computer. The fewer finished computers Dell holds in inventory,

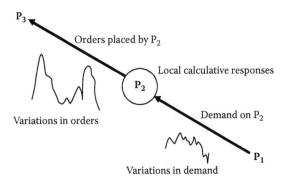

FIGURE 1.2
Variation amplification caused by the bullwhip effect.

the less money they lose per computer as they "rot" on a shelf. The result is near-perfect visibility of customer demand and inventory movement throughout the supply chain including manufacturing process. Better information leads to better inventory positioning and lower costs throughout the supply chain.

Barriers to the implementation of a demand-driven manufacturing include the necessary investment in the three toolboxes described in the next section.

1.3 Three Toolboxes for Lean Manufacturing

In a nutshell, Lean manufacturing means "manufacturing without waste." Many Lean tools are available to assist companies in their Lean manufacturing journey. These tools constitute a toolbox that helps to eliminate waste in every area of production including customer relations, product design, supplier networks, and factory management. As described in the following, this toolbox enables companies to incorporate less human effort, less inventory, less time to develop products, and less space to become highly responsive to customer demand while producing top-quality products in the most efficient and economical manner possible. According to the tools' alphabetic order, we are grouping into three toolboxes.

1.3.1 Toolbox 1: From 5S to Kanban

Table 1.1 summarizes the Lean manufacturing tools from 5S to Kanban. Here are detailed discussions on a few tools.

1.3.1.1 The 5S System

The 5S system is designed to improve workplace organization and standardization. The Ss are:

1. Sort—Sort through all items and remove nonvalue-added items.
2. Straighten—Set in order remaining items, set limits, and create temporary location indicators.
3. Shine—Clean everything and use cleaning as inspection.
4. Standardize—Standardize the first three Ss by implementing visual displays and controls.
5. Sustain—Sustain the gains through self-discipline, training, communication, and total employee involvement.

TABLE 1.1

Description of Lean Manufacturing Tools: From 5S to Kanban

Tool	About the Tool
5S	A methodology for organizing, cleaning, developing, and sustaining a productive work environment. Improved safety, ownership of workspace, improved productivity, and improved maintenance are some of the benefits of a 5S program.
Error proofing	A structured approach to ensure quality and an error-free manufacturing environment. Error proofing assures that defects will never be passed to next operation.
Current reality trees	A problem-analysis tool, aids to examine cause and effect logic behind our current situation.
Conflict resolution diagram	Used to resolve hidden conflicts that usually perpetuate chronic problems.
Future reality diagram	A sufficiency-based logic structure designed to reveal how changes to the status quo would affect reality, specifically to produce desired effects.
Inventory turnover rate	The number of times an inventory cycles or turns over during the year. A frequently used method to compute inventory turnover is to divide average inventory level into annual cost of sales.
JIT	A philosophy of manufacturing based on planned elimination of all waste and continuous improvement of productivity. It encompasses the successful execution of all manufacturing activities required to produce a final product.
Kaizen	The Japanese term for improvement; continuing improvement involving everyone—managers and workers. In manufacturing, kaizen relates to finding and eliminating waste in machinery, labor, or production methods.
Kanban	Kanban is a simple parts-movement system that depends on cards and boxes or containers to take parts from one workstation to another on a production line. The essence of the kanban concept is that a supplier or the warehouse should only deliver components to the production line as and when they are needed, so that there is no storage in the production area.

Another way to summarize 5S is a place for everything (first 3Ss) and everything in its place (last two Ss).

1.3.1.2 Cellular Manufacturing

In cellular manufacturing, workstations and equipment are arranged in a sequence that supports a smooth flow of materials and components through the production process with minimal transport or delay. When processes are balanced, the product flows continuously and customer demands are easily met. Cellular/flow manufacturing is the linking of manual and machine operations into the most efficient combination of resources to maximize value-added content while minimizing waste.

Rather than processing multiple parts before sending them on to the next machine or process step, cellular manufacturing moves products through the manufacturing process one piece at a time, at a rate determined by customers' needs. Cellular manufacturing also provides companies the flexibility to vary product type or features on the production line in response to specific customer demands. This manufacturing approach seeks to minimize the time it takes for a single product to flow through the entire production process. As a result, parts movement is minimized, wait time between operations is reduced, inventory is reduced, and productivity increases.

1.3.1.3 Kaizen

Kaizen is a system of continuous improvement in quality, technology, processes, company culture, productivity, safety, and leadership. The word *kaizen* means "continuous improvement" in Japanese. Kaizen is an intensive and focused approach to process improvement. This continuous improvement methodology combines Lean manufacturing tools such as the 5Ss of workplace organization and standardization, cells, pull/Kanban, setup reduction, and line balancing. Each tool incorporates team empowerment, brainstorming, and problem solving to rapidly make improvements to a specific product or portions of your processes.

Kaizen strategy calls for never-ending efforts for improvement involving everyone in the organization—managers and workers alike. The kaizen methodology has been used extensively for improving the organization of work in factories and actual methods used to manufacture products. The results are real-time with implementation occurring within one week. Not only will you see immediate improvements to your process—you will also develop a list of the improvement opportunities that your staff can investigate and implement after the kaizen. Kaizen will provide your company with immediate tangible results, motivation, and ongoing continuous improvement within your company.

1.3.1.4 Kanban

Kanban is a signaling system to trigger action. As its name suggests, Kanban historically uses cards to signal the need for an item. However, other devices such as plastic markers (kanban squares) or balls (often golf balls) or an empty-part transport trolley or floor location can also be used to trigger the movement, production, or supply of a unit in a factory.

Kanban controls the flow of resources in a production process by replacing only what have been consumed. They are customer order-driven production schedules based on actual demand and consumption rather than forecasting. Implementing Kanban can help you eliminate waste in handling, storing, and getting your product to the customer on time, every time.

1.3.2 Toolbox 2: From Lean Metrics to Standard Rate

Table 1.2 summarizes the Lean manufacturing tools from Lean metrics to standard rate.

Here are detailed discussions on a few tools.

1.3.2.1 Lean Office

Lean office is the application of the Lean manufacturing philosophy to the office and administrative processes. Lean office helps companies with any type of administrative function to streamline information flow—the gathering, improving, movement, and storage of information. Just like Lean for the factory floor, Lean office focuses on reducing total cycle time—in this case, the time between orders being placed and when payments are received.

Over 40 percent of total cycle time occurs at the front end of a project. Tasks such as taking orders, confirming credit, designing parts, and ordering materials eat up approximately 42 percent of a typical company's total cycle time. Manufacturing and shipping account for only 8 percent. The remaining 50 percent of the cycle time is spent waiting for customer payment. One of the important aspects of Lean is the focus on system-level improvements rather than point improvements. Within this system level of thinking, office work and shop-floor work often converge. Work in either the office or on the shop floor can have significant impact on customer satisfaction. This knowledge, combined with the understanding of how to remove waste properly, is critical for any successful Lean manufacturing.

1.3.2.2 Single Minute Exchange of Die (SMED)

SMED, which stands for single minute exchange of die, is a systematic way to reduce or ultimately eliminate the time it takes to set up or change over for a manufacturing job. The SMED method was developed by Shigeo Shingo to dramatically reduce or eliminate changeover time. The method is used to help companies design no- or low-cost solutions to reduce changeover time. This, in turn, allows the company to meet customer demands for high-quality, low-cost products, delivered quickly and without the expense of excess inventory.

SMED reduces changeover times to less than ten minutes. By reducing setup time toward SMED, more productions can be completed each day. This enables smaller batch sizes to be produced while reducing lead time.

1.3.3 Toolbox 3: From Takt Time to Workflow Diagram

Table 1.3 summarizes the Lean manufacturing tools takt time to workflow diagram. Following are detailed descriptions on a few tools.

TABLE 1.2

Description of Lean Manufacturing Tools: From Lean Metrics to Standard Rate

Tool	About the Tool
Lean metrics	Lean metrics allow companies to measure, evaluate, and respond to their performance in a balanced way, without sacrificing the quality to meet quantity objectives, or increasing inventory levels to achieve machine efficiencies. The type of the Lean metric depends on the organization and can be of the following categories: financial performance, behavioral performance, and core process performance.
Lean office	Helps companies with any type of administrative function to streamline information flow—the gathering, improving, movement, and storage of information.
Lean performance indicator (LPI)[a]	A consistent method to measure Lean implementation effectiveness. A key core value metric for motivating performance and rewarding team performance through the PIP plus incentive program. Indicators: Real-time performance, continuous improvement implementation, Lean sustainment, waste elimination, and profitability. Goal: An LPI monthly goal of 100; equates to 116.3 percent value-added output performance at level C Lean performance. Formula: Value-added sales (total sales minus raw materials, subcontracting, and components) divided by shop rate per hour () divided by number of hourly shop floor personnel divided by 2.
One-piece flow	A processing concept in which items are processed and moved directly from one processing step to the next, one piece at a time. One-piece flow helps to maximum utilization of resources, shorten lead times, identify problems, and communication between operations. Also called continuous flow.
Overall equipment effectiveness (OEE)	Measures the availability, performance efficiency, and quality rate of equipment; it is especially important to calculate OEE for the constrained operations.
Prerequisite tree	A logical structure designed to identify all obstacles and the responses needed to overcome them in realizing an objective. It identifies minimum necessary conditions without which the objective cannot be met.
Process route table	Shows what machines and equipment are needed for processing a component or assembly. These tables aid in creating ordinary lines and grouping work pieces into workcells.
Quick changeover	A technique to analyze and reduce resources needed for equipment setup, including exchange of tools and dies. Single minute exchange of dies (SMED) is an approach to reduce output and quality losses due to changeovers.
Standard rate or work	The length of time that should be required to set up a given machine or operation and run one part, assembly, batch, or end product through that operation. This time is used in determining machine requirements and labor requirements.

[a] *Source:* Warren, J. LPI, Copyright by ShopWerks Software, 2004–2006.

TABLE 1.3

Description of Lean Manufacturing Tools: From Takt Time to Workflow Diagram

Takt time	The time required between completion of successive units of end product. Takt time is used to pace lines in the production environments.
Theory of constraints (TOC)	A management philosophy that can be viewed as three separate but interrelated areas: logistics, performance measurement, and logical thinking. TOC focuses the organization's scarce resources on improving the performance of the true constraint, and therefore the bottom line of the organization.
Total productive maintenance (TPM)	A maintenance program concept that brings maintenance into focus in order to minimize downtimes and maximize equipment usage. The goal of TPM is to avoid emergency repairs and keep unscheduled maintenance to a minimum.
Toyota Production System (TPS)	A technology of comprehensive production management. The basic idea of this system is to maintain a continuous flow of products in factories to flexibly adapt to demand changes. The realization of such production flow is called just-in-time production, which means producing only necessary units in a necessary quantity at a necessary time. As a result, the excess inventories and the excess workforce will be naturally diminished, thereby achieving the purposes of increased productivity and cost reduction.
Transition tree	A cause-and-effect logic tree designed to provide step-by-step progress from initiation to completion of a course of action or change. It is an implementation tool.
Value-added to nonvalue-added lead-time ratio	Provides insight on how many value-added activities are performed compared to nonvalue-added activities, using time as a unit of measure.
Value stream costing	A methodology that simplifies the accounting process to give everyone real information in a basic understandable format. By isolating all fixed costs along with direct labor we can easily apply manufacturing resources as a value per square footage utilized by a particular cell or value stream. This methodology of factoring gives a true picture of cellular consumption to value-added throughput for each value stream company-wide. Now you can easily focus improvement kaizen events where actual problems exist for faster calculated benefits and sustainability.
Value stream mapping	A graphical tool that helps you to see and understand the flow of the material and information as a product makes its way through the value stream. It ties together Lean concepts and techniques.
Visual management	A set of techniques that makes operation standards visible so that workers can follow them more easily. These techniques expose waste so that they can be prevented and eliminated.
Workflow diagram	Shows the movement of material, identifying areas of waste. Aids teams to plan future improvements, such as one-piece flow and workcells.

1.3.3.1   *Total Productive Maintenance*

Total productive maintenance (TPM) is a process to maximize the productivity of plant equipment for their entire life. Considered the medical science of machines, TPM reduces equipment downtime while improving quality and capacity. Involving a newly defined concept for maintaining plants and equipment, TPM fosters an environment where improvement efforts in safety, quality, delivery, cost, and creativity are encouraged through the participation of all employees.

TPM brings maintenance into focus as a necessary and vitally important part of the manufacturing. It is no longer regarded as a nonvalue-added activity. Downtime for maintenance is scheduled as a part of the manufacturing day and, in some cases, as an integral part of the manufacturing process. It is no longer simply squeezed in whenever there is a break in material flow. TPM is a proactive approach that minimizes emergency and unscheduled maintenance.

1.3.3.2 Value Stream Mapping

Value stream mapping (VSM) is a tool used to create a material and information flow map of a product or service. VSM identifies waste within a process. It focuses resources on issues that will make the most significant process improvements.

VSM allows companies to map the flow of materials and information from order to cash as well as throughout the supply chain. Mapping the value stream, you begin the journey with the current state map—it shows you where you are. Then, you plan your Lean journey with a future state map—it shows you where you are going and how you are going to get there. Based on your value stream map, you can streamline work processes, thereby, cutting lead times and reducing operating costs. The VSM quickly defines the sequential process steps and data pertinent to each of the steps as well as overall metrics relating to the entire process. The map also shows how information flows, where rework occurs and where there may be quality issues within the process. Value stream maps generally provide broad overview of the process at a high level. VSM can be used to identify potential bottlenecks and opportunities for improvement.

1.4 Let's Play a "Lean" Game: Manufacturing a Gyrocopter

This Lean game transforms the famous beer distribution game into a gyrocopter manufacturing game by integrating the following two sources:

- Beer distribution game (beer game)—This is a simulation game created by a group of professors at the MIT Sloan School of Management in the early 1960s to demonstrate a number of key principles of supply chain management. The game is played by teams of at least four players, often in heated competition, and takes from one to one and a half hours to complete. A debriefing session of roughly equivalent length typically follows to review the results of each team and discuss the lessons involved. The purpose of the game is to meet customer demand for cases of beer through a multistage supply chain with minimal expenditure on back orders and inventory.

- Gyrocopter robust design game—This robust design game was published in a book titled *Engineering Robust Design with Six Sigma* (Wang, 2005). This game was organized to show how to identify control factors and noise factors in design and manufacturing. This game is played by a Lean Six Sigma engineer or a Lean Six Sigma product development team. After completing the robust design, a presentation usually follows to review the results of each gyrocopter design in terms of the engineering scorecards for Lean Six Sigma. The purpose of the game is to meet customer demand for cases of gyrocopters through a multistage robust design with minimal variation on critical-to-quality (CTQ) characteristics.

1.4.1   Materials, Tools, and Steps for Manufacturing the Gyrocopter

The following materials and tools are required for manufacturing the gyrocopter:

- Paper
- Ruler
- Scissors or pinking shears
- Paper clip

The steps for manufacturing the gyrocopter are as follows:

1. Patterning—Cut out a 6 1/2-inch long by 1 1/2-inch wide strip of paper (see Figure 1.3).
2. Creating wing—Starting at the top, cut a 3-inch slit down through the middle of the strip to create a pair of wings (see Figure 1.4).
3. Final assembly—Fold down the wings in opposite directions from one another. Attach a paper clip to the bottom of the strip for weight (see Figure 1.5).

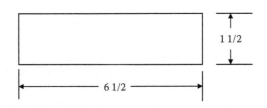

FIGURE 1.3
Paper patterning for gyrocopter.

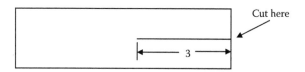

FIGURE 1.4
Creating wings for gyrocopter.

Now you can perform a quality control by dropping the finished gyrocopter from an elevated spot, and it should spin to the ground.

1.4.2 Establish an Assembly Line for Manufacturing Gyrocopters

Each week, customers purchase gyrocopters from the gyrocopter store, which provides the gyrocopters requested out of inventory. The gyrocopter store in turn places orders for more gyrocopters with the Assembly Department, which ships the assembled gyrocopters requested out of its own inventory. The Assembly Department orders and receives wings from the Wing Department, who in turn orders and receives patterns from the Patterning Department, where the paper is patterned. At each stage there are transporting delays and order-processing delays.

As shown in Figure 1.6, the numbers in the boxes show typical initial conditions. The simulation begins in an equilibrium where each player has an inventory of twelve gyrocopters, and orders, shipments, and gyrocopters in the pipeline of order processing and shipping delays all reflect a steady throughput of gyrocopters per week. In the classic game consumer demand begins at four gyrocopters per week (for the first four weeks), then jumps to eight cases of gyrocopters per week and remains there for the remainder of the game. Alternative demand patterns will be discussed later.

The boxes with numbers represent the delays in processing gyrocopter orders. When a player places an order, it takes two weeks for the order to be received by their upstream supplier. The boxes labeled with "shipping delay"

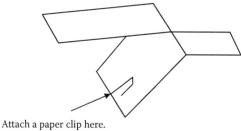

Attach a paper clip here.

FIGURE 1.5
Gyrocopter final assembly.

Time	Actual Order by Gyrocopter Store	Perceived Demand by Assembly Department	Perceived Demand by Wing Department	Perceived Demand by Patterning Department
1	503	10068	10102	10153
2	520	10397	10598	10901
3	503	10064	10096	10145
4	483	9654	9482	9227
5	474	9488	9235	8860
6	508	10155	10233	10350
7	539	10775	11170	11771
8	556	11116	11689	12567
9	514	10272	10409	10616
10	503	10059	10089	10133
11	544	10871	11316	11994
12	499	9988	9982	9973
13	516	10321	10483	10728
14	494	9885	9828	9742
15	462	9245	8875	8329
16	511	10216	10324	10488
17	499	9988	9982	9974
18	540	10807	11219	11845
19	494	9878	9818	9727
20	512	10247	10371	10558
21	515	10309	10465	10701
22	548	10950	11436	12179
23	516	10314	10472	10710
24	491	9822	9734	9602
25	526	10529	10797	11203
26	507	10134	10201	10302
27	501	10015	10022	10033
28	495	9908	9862	9793
29	489	9772	9658	9489
30	498	9959	9938	9907

FIGURE 1.6
Establish an assembly line for manufacturing gyrocopters.

and "transporting delay" represent the delays in shipping and transporting materials, work in process (WIP), and finished gyrocopters. When players send off a shipment, it takes two weeks for the shipment to be received by their downstream customer. Note that for the manufacturer, it takes three weeks to manufacture a gyrocopter. Cost functions on these delays are discussed in Section 1.3.

1.4.3 Develop Cost Functions for Manufacturing Gyrocopters

The players' objective is to minimize total team costs. It costs $.50 for each gyrocopter or WIP that each player holds in inventory each week. If your customer has ordered gyrocopters or WIPs and you have none in inventory, you incur a backlog cost of $1/case/week, to capture both the lost revenue and the ill will a stockout causes between the upstream customer and the downstream supplier. Furthermore, backlogged orders build up; if you have not served a customer in week 3, you can sell to them from gyrocopters that arrive at your site in week 4. Costs are assessed at each link of the factory chain.

Information management impacts factory cost functions significantly. Each player has good local information about her or his factory only. Players keep records of their inventory, backlog, and orders placed with their supplier each week. There is no communication (other than the transmittal of orders and shipments) allowed between positions between factories. Customer demand is not known to any of the players in advance. Only the gyrocopter store discovers customer demand as the game proceeds. The others learn only what their own upstream customer orders. This restriction means that the players can't explicitly coordinate their decisions when placing orders, even though the objective of each team is to minimize total costs.

1.4.4 Ground Rules and Course for Playing the Lean Manufacturing Game

1.4.4.1 Ground Rules

A game extends over a fictitious year and covers 52 rounds of one week each. Each round takes 60 to 120 seconds and thus the total playing time is 60 to 120 minutes. The entire game must be played in one go. In other words, you cannot take a break.

For the sake of simplicity, everyone sells only one product: G101 Gyrocopter. One unit equals one gyrocopter (finished good or WIP).

There are two costs involved in the game: inventory carrying costs and backlog costs. The inventory carrying cost per unit (Gyrocopter – Finished good or WIP) in stock is $1 per week. The backlog cost per unit that you fail to deliver is $2 per week.

In order to reflect reality better, two delays are introduced between you and your fellow players:

1. From when you reach a decision, it takes two weeks before your supplier receives the order.
2. It takes two weeks before his delivery reaches you.

1.4.4.2 Course of the Game

Here's how one week of the game works (the five levels of the game):

1. Customers arrive at the Gyrocopter Store to buy gyrocopters.
2. The Gyrocopter Store:
   a. Receives a shipment from the Assembly Department (that had been shipped two weeks before).
   b. Sells gyrocopters to customers, filling all orders, if possible; recording backlogs, if necessary.
   c. Orders more gyrocopter wings from the Assembly Department to replenish inventory, if necessary.
3. The Assembly Department:
   a. Receives shipment from the Wing Department (that had been shipped two weeks before).
   b. Receives an order from the Gyrocopter Store (that had been sent two weeks before).
   c. Ships gyrocopters to the Gyrocopter Store, filling the order, if possible; recording backlogs, if necessary.
   d. Orders more gyrocopter wings from the Wing Department to replenish inventory, if necessary.
4. The Wing Department:
   a. Receives shipment from the Patterning Department (that had been shipped two weeks before).
   b. Receives an order from the Assembly Department (that had been sent two weeks before).
   c. Ships gyrocopters to the Assembly Department, filling the order, if possible; recording backlogs, if necessary.
   d. Orders more gyrocopter patterns from the Patterning Department to replenish inventory, if necessary.
5. The Patterning Department:
   a. Receives shipment of raw materials (paper and paper clips) from the supplier's facility (whose production had been scheduled two weeks before).
   b. Receives an order from the Wing Department (that had been sent two weeks before).
   c. Ships gyrocopter patterns to the Wing Department, filling the order, if possible; recording backlogs, if necessary.
   d. Schedules the production of more gyrocopters to replenish inventory, if necessary.

During the course of the game, players must not exchange any information other than that constituted by the order itself.

1.4.4.3   *End the Game*

The game ends after 52 weeks, but can be concluded at any time before that by the game leader.

1.4.5   Bullwhip Effect Typical Results for Playing the Lean Manufacturing Game

The gyrocopter game is deceptively simple compared to real manufacturing business. All you have to do is meet customer demand and order enough from your downstream supplier to keep your inventory low while avoiding costly backlogs. There are no machine breakdowns or other random events, no labor problems, no capacity limits or financial constraints. Yet the results are shocking. Average team costs are about $2,000, though it is not uncommon for costs to exceed $10,000; few ever go below $1,000. Optimal performance (calculated using only the information actually available to players themselves) is about $200. Average costs are ten times greater than optimal!

The results are shown in figures and diagrams when the game is over. Figure 1.7 shows that the demand by assembly line position varies dramatically even for constant store orders.

As shown in Figure 1.8, the departures from optimality are not random. Though individual games differ quantitatively, they always exhibit the same patterns of behavior:

- Oscillation—Orders and inventories are dominated by large amplitude fluctuations, with an average period of about 20 weeks.
- Amplification—The amplitude and variance of orders increases steadily from the Gyrocopter Store to the Patterning Department. The peak order rate at the Patterning Department is on average more than double the peak order rate at the Gyrocopter Store.
- Phase lag—The order rate tends to peak later as one moves from the Gyrocopter Store to the Patterning Department.

In virtually all cases, the inventory levels of the retailer decline, followed in sequence by a decline in the inventory of the Assembly Department, Wing Department, and Patterning Department. As inventory falls, players tend to increase their orders. Players soon stock out. Backlogs of unfilled orders grow. Faced with rising orders and large backlogs, players dramatically boost the orders they place with their upstream supplier. Eventually, the factory patterns and ships this huge quantity of WIPs, and inventory levels surge. In many cases one can observe a second cycle.

Time	Actual Order by Gyrocopter Store	Perceived Demand by Assembly Department	Perceived Demand by Wing Department	Perceived Demand by Patterning Department
1	503	10068	10102	10153
2	520	10397	10598	10901
3	503	10064	10096	10145
4	483	9654	9482	9227
5	474	9488	9235	8860
6	508	10155	10233	10350
7	539	10775	11170	11771
8	556	11116	11689	12567
9	514	10272	10409	10616
10	503	10059	10089	10133
11	544	10871	11316	11994
12	499	9988	9982	9973
13	516	10321	10483	10728
14	494	9885	9828	9742
15	462	9245	8875	8329
16	511	10216	10324	10488
17	499	9988	9982	9974
18	540	10807	11219	11845
19	494	9878	9818	9727
20	512	10247	10371	10558
21	515	10309	10465	10701
22	548	10950	11436	12179
23	516	10314	10472	10710
24	491	9822	9734	9602
25	526	10529	10797	11203
26	507	10134	10201	10302
27	501	10015	10022	10033
28	495	9908	9862	9793
29	489	9772	9658	9489
30	498	9959	9938	9907

FIGURE 1.7
Demand by assembly line position in response to constant store orders.

The bullwhip effect has become popular through the observation of Procter & Gamble (P&G) in the supply chain of its Pampers diapers. P&G first coined the term *bullwhip effect* to describe the systematic ordering behavior witnessed between customers and suppliers of Pampers diapers (Lee, Padmanabhan, and Whang, 1997). Babies use diapers at a very predictable rate, and retail sales resemble this fact. Information is readily available concerning the number of babies in all stages of diaper wearing. Even so P&G observed that this product with uniform demand created

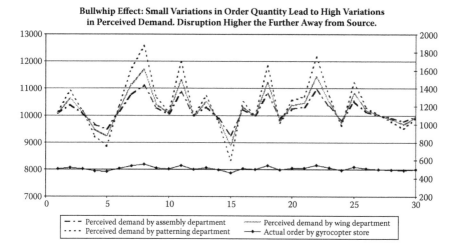

FIGURE 1.8
Variation with store orders being amplified along assembly line.

a wave of changes up the supply chain due to very minor changes in demand.

In the early 1990s P&G faced a problem of extreme demand variations for Pampers diapers. The logistics executives at P&G examined the order rates for Pampers across the supply chain. While customers use diapers at a fairly constant rate, the logistics executives found that wholesale orders fluctuated considerably over time. The firm also found that the orders it placed for raw materials with its suppliers fluctuated even more than these wholesale orders. This classic example stimulates the rapid deployment of Lean manufacturing in major corporations.

1.5 Five Stages of Lean Manufacturing: Business Bottom-Line Based

As shown by the gyrocopter game, the bullwhip effect is the magnification of demand fluctuations, not the magnification of demand. The bullwhip effect is evident in an assembly/manufacturing line when demand increases and decreases. The effect is that these increases and decreases are exaggerated down the assembly/manufacturing line.

Some of the reasons that the bullwhip effect occurs include the following:

- Overreacting to the backlog orders
- Little or no communication between supply chain partners

- Delay times between order processing, demand, and receipt of products
- Order batching, method for reduction of ordering costs due to price discounts for bulk ordering, transportation expense decrease by ordering full-truck loads, and so forth
- Limitations on order size (i.e., retailers can order products in cases of ten from wholesaler; however, distributors receive orders in cases of 1,000)
- Inaccurate demand forecasts

Excess raw materials costs arise from the last-minute purchasing decisions made to accommodate an unplanned increase in demand. The result of these panicked buying periods is an inventory of unused supplies. As these unused supplies grow, so do the associated costs.

Inefficient utilization and overtime expenses incurred during high demand periods follow excess capacity during periods of low volume of demand. This is made worse by the excess warehousing expenses that are incurred because of unused storage space as well as increases in shipping costs caused by premium rates paid for last minute orders.

When the bullwhip effect is first identified in an assembly/manufacturing line, it is important to identify the problem areas. Once changes are made in these areas, the productivity and timeliness of the supply chain will increase greatly and the bullwhip effect will be dramatically lessened as shown by the following example.

EXAMPLE 1.1

Hewlett Packard sells its printers in almost every country. It can forecast overall demand for its printers, but trying to pinpoint what country they will be sold in is very difficult. Because different countries employ different power requirements and the language varies (pertinent to user manuals), Hewlett Packard redesigned its printer so that it could wait until demand materializes and then add these country-specific items at its regional distribution centers rather than at central factories.

The following five-step Lean manufacturing process can be utilized to minimize the bullwhip effect.

1. Specify value—Define value from the perspective of the final customer. Express value in terms of a specific product, which meets the customer's needs at a specific price and at a specific time.
2. Map value stream—Identify the value stream, the set of all specific actions required to bring a specific product through the three critical management tasks of any business:
   - The problem-solving task
   - The information management task
   - The physical transformation task

Create a map of the current state and the future state of the value stream. Identify and categorize waste in the current state, then eliminate it.

3. Streamline production flow—Make the remaining steps in the value stream flow. Eliminate functional barriers and develop a product-focused organization that dramatically improves lead time.

4. Let customers pull products—Let the customer pull products as needed, eliminating the need for a sales forecast.

5. Energize continuous improvement—As shown by Figure 1.8, there is no end to the process of reducing effort, time, space, cost, and mistakes. Return to the first step and begin the next Lean transformation, offering a product that is what the customer wants.

In the following chapters, we will show how to use this five-step process to tame the bullwhip effect and other manufacturing risks.

2

Put Business Bottom Line First:
Transfer Function for Production Cost

This chapter enables you to

- Understand the transfer function of production cost and analyze how companies produce and offer goods for sales
- Recognize the difference between short-term and long-term cost functions, marginal cost function, and the shapes of various cost curves
- Analyze your industrial sectors to prevent production failures and win in the market

2.1 Production Transfer Function

Lean manufacturing is the production of goods using less of everything compared to mass production: less human effort, less manufacturing space, less investment in tools, and less engineering time to develop a new product. In Lean manufacturing, the production transfer function asserts that the maximum production of a technologically determined production process is a mathematical function of input factors of production. Considering the set of all technically feasible combinations of output and inputs, only the combinations encompassing a maximum output for a specified set of inputs would constitute the production function. Alternatively, a production function can be defined as the specification of the minimum input requirements needed to produce designated quantities of output, given available technology. It is usually presumed that unique production functions can be constructed for every production technology.

The transfer function for production cost specifies the maximum output that can be produced with a given quantity of inputs for a given state of engineering and technical knowledge. The production function relates the production to the amount of inputs, typically capital and labor. It is important to keep in mind that the production transfer function describes technology, not economic behavior. A company may maximize its profits given its production transfer function, yet generally take the production function as a given element of that problem. In specialized long-term models, the company may choose its capital investments among production technologies.

2.2   Short-Term vs. Long-Term Production Transfer Functions

Lean is about doing more with less: less time, inventory, space, labor, and money. Lean manufacturing is shorthand for a commitment to eliminating waste, simplifying procedures, and speeding up production. The distinction between the short term and long term is important in Lean production because each period has its own transfer function for production cost.

2.2.1   Lean Manufacturing: Optimize Short-Term Production Transfer Function

The short term is the period in which companies can adjust production only by changing variables factors, such as materials and labor; however fixed factors, such as capital, cannot change. Lean manufacturing (also known as the Toyota Production System) is, in its most basic form, the systematic elimination of waste—overproduction, waiting, transportation, inventory, motion, overprocessing, defective units—and the implementation of the concepts of continuous flow and customer pull.

Lean manufacturing is a proven approach to reduce waste and streamline operations. Lean manufacturing embraces a philosophy of continually increasing the proportion of value-added activity through ongoing waste elimination. A Lean manufacturing approach provides companies with tools to survive in a global market that demands higher quality, faster delivery, and lower prices. Specifically:

- Lean manufacturing dramatically reduces the waste chain.
- Lean manufacturing reduces inventory and floor space requirements.
- Lean manufacturing creates more robust production systems.
- Lean manufacturing develops appropriate material delivery systems.
- Lean manufacturing improves layouts for increased flexibility.

2.2.2   Six Sigma: Optimize Long-Term Production Transfer Function

The long term is a period sufficiently long enough so that all factors in the production function, including capital, can be adjusted. Six Sigma is a philosophy of doing business with a focus on eliminating defects through fundamental process knowledge. Six Sigma methods integrate principles of business, statistics, and engineering to achieve tangible results.

Six Sigma tools are used to improve the processes and products of a company. They are applicable across every discipline, including production, sales, marketing, design, administration, and service. Six Sigma offers a wealth of tangible benefits. When skillfully applied:

- Six Sigma reduces costs by 50 percent or more through a self-funded approach to improvement.
- Six Sigma reduces the waste chain.
- Six Sigma affords a better understanding of customer requirements.
- Six Sigma improves delivery and quality performance.
- Six Sigma provides critical process inputs needed to respond to changing customer requirements.
- Six Sigma develops robust products and processes.
- Six Sigma drives improvements rapidly with internal resources.

Lean manufacturing and Six Sigma are toolkits to reduce waste in business processes. Both Lean manufacturing and Six Sigma are proven concepts and have saved millions of dollars for the world's manufacturers.

2.3   Short-Term Transfer Function for Production

The importance of Lean manufacturing is better comprehended when its impact of change on economics is thoroughly understood. The manufacturing engineering philosophy is pivoted on designing a manufacturing system that perfectly blends together the fundamentals of minimizing cost and maximizing profit.

The production's *fixed costs*, sometimes called "overhead," are those costs that do not change with production or sales levels. Examples of fixed costs include rent, property tax, insurance, interest expense, and salaries of exempt employees. Automobile repairs now shift focus to "fixed" costs—repairing more cars with the same fixed cost or repairing the same amount of cars with a lower fixed cost.

Variable costs are those costs that change with the level of output. For example, when you increase production to meet demand, you have to pay for more raw materials and fuel. You also have to pay more in wages to cover the increased overtime and additional workers. Unless you are going to eliminate some fixed costs, the only real cost reduction is the variable cost. If the supplier cannot produce the part for a price lower than your variable cost, you are not saving your company money.

In the most simple production function, *total cost* is equal to fixed costs plus variable costs. For Lean manufacturing, the difference between fixed costs and variable costs is crucial, as each will influence production decisions for profit maximization differently. The only meaningful measurement of total cost is on a cash basis. All money spent on manufacturing must be summarized and the total compared to the previous period, not to a flexible budget or a plan. What matters is whether the total cash spent on manufacturing was more or less than it was in the previous period.

It is important that this cost figure is exclusive of all allocations, and does not exclude sales, general, and administrative expense. The only exceptions are that major capital investment spending is excluded, and expenses are adjusted for accounts receivable and payable. While these amounts must be added back in to create a total Lean-accounting-based income statement, manufacturing performance should be measured as if payment were made at the time materials and services were delivered, and payment were collected at the time finished goods were shipped to an outside customer.

Consider the following example. XYZ Company produces many of their machined parts in-house. XYZ's materials manager has been tasked with reducing product cost. He begins with part A since it is the highest volume part and has the highest potential for cost reduction. Part A has a standard cost of $2,300. The variable-cost component (labor and material) is $1,200. The fixed-cost component (overhead including machine tools, building, etc.) is $1,100. A supplier quotes the part at $1,800. Because XYZ produces 3000 of these per year, the materials manager believes he is saving the company $1.5 million per year by outsourcing this part. This apparent savings is shown in Table 2.1, yet the savings are not real.

Here, the only real cost reduction is the variable cost reduction in this case. The company has reduced its cost by $1,200 per part yet the savings is offset by a $1,800 per part cost to the supplier. XYZ is operating at a higher cost by outsourcing this part. XYZ has increased its costs by $1.8 million per year; however, in XYZ's enterprise resource planning (ERP) system, the standard cost of part A is now lower ($1,800 vs. $2,300). As shown by Table 2.2, the fact that costs have increased shows itself in the form of higher standard costs for all other

TABLE 2.1

Cost Saving Is Apparent, Yet Not Real

Standard Costs for Part A	In-House	After Outsourcing
Labor	$1,100	
Materials	$100	$1,800
Fixed costs (overhead)	$1,100	
Total (standard cost)	$2,300	$1,800
Apparent savings per part		$500
Apparent annual cost reduction (3000/year)		$1,500,000

TABLE 2.2

Total Cost: Reflects Company's Bottom Line

Relevant Costs Annualized	In-House	After Outsourcing
Labor	$3,300,000	—
Materials	$300,000	$5,400,000
Fixed costs (overhead)	$3,300,000	$3,300,000
Total	$6,900,000	$8,700,000
Actual additional annual cost		$1,800,000

Note: Fixed costs would be allocated to other parts after outsourcing.

parts using the same facilities as part A. This gives the materials manager incentive to outsource more parts thus creating a cycle of increasing standard costs for in-house parts and savings on paper by outsourcing. The underlying reality of higher costs is not obvious except in the company's bottom line.

To determine if lower costs are possible by outsourcing, one must consider all relevant costs in detail. Factors that will influence whether lower costs can be achieved by outsourcing are process and labor costs. If a supplier has a better process for producing the part, it is possible that he can produce it for less than your organization. Also, if a supplier has lower labor costs (such as a supplier in Mexico or China), again it would be possible that he can produce the part at a lower cost. If neither of these is true, one must be skeptical that a lower cost can be achieved by outsourcing. The supplier must have some profit built into its price. Given that fact, if they are using a similar process and are working with the same labor pool that you are, it is unlikely that you will save your organization money simply by outsourcing parts currently manufactured in-house.

2.4   Transfer Function for Marginal Cost

Marginal cost is the additional cost incurred in producing one extra unit of a product. It is arguably the most important kind of cost. In the simplest cases, fixed costs do not affect production decisions, because they cannot be changed, and management will choose to produce if sales prices are above the cost of each additional unit (marginal cost).

Mathematically, the marginal cost (MC) function is expressed as the derivative of the total cost (TC) function with respect to quantity (Q). Note that the marginal cost may change with volume, and so at each level of production, the marginal cost is the cost of the next unit produced.

$$MC = \frac{dTC}{dQ}$$

(2.1)

In general terms, marginal cost at each level of production includes any additional costs required to produce the next unit. If producing additional vehicles requires, for example, building a new factory, the marginal cost of those extra vehicles includes the cost of the new factory. In practice, the analysis is segregated into short- and long-run cases, and over the longest run, all costs are marginal. At each level of production and time period being considered, marginal costs include all costs that vary with the level of production, and other costs are considered fixed costs.

When one player in a particular industry sector begins to lower marginal costs through supply chain efficiencies, the others follow suit to remain competitive from the marginal cost point of view. Despite this economic expectation, variations across companies in a common sector are very common. A good example is a comparison of Dell and HP where Dell carries a five-day inventory as compared to HP's twenty-five. Experts estimate that a week's inventory advantage translates into a 1 percent cost advantage, which contributes to lowering Dell's marginal cost and contributes to profitability in a cutthroat price competition. Depending on a company's enterprise to gain supply chain advantage, the contingency plans in supply chains for September 11–like interruptions have varied across the industry. Whereas some have worked on innovative ideas to keep a Lean inventory flow in interruptive circumstances, others that have demonstrated limited interest in supply chain reengineering do not seem to have any concrete contingency plans. One theory that may explain this variance is that indifferent companies often operate with high inventory levels and low turnaround frequencies, which makes them less vulnerable during crises. That being said, such crises contribute to economic downturns, which can have a negative impact on firms with inefficient inventory management in the long run. While the September 11 episode has forced experts to look at contingency measures in the form of increased inventory levels, this philosophy needs to be applied as an exception than a rule given that there is little denial on the payoffs gained in minimizing inventory in supply chains and improving turnaround. Although a management emphasis has always been useful to ring in a supply chain focus, companies also need to look beyond the first tier suppliers to make sure that all constituents of the chain comply with adopting positive measures in supply chain management. This is because all companies in a chain or a network are vulnerable to crisis circumstances even if one constituent of the chain faces problems. The other area that can greatly contribute to improving efficiencies is the adeptness to measure all inventory and information flows in the chain. The lack of data is one of the key issues in identifying problems and recognizing required measures. Although the purchase of software for supply chain management may signal instituted measures including one for data collection, organizational structure and dedicated experts are key to intelligent interpretation of data and optimization of software functions.

2.5 The Law of Diminishing Returns: Key to Understanding Lean Manufacturing

For traditional break-even point analysis of production, it is assumed that total costs and revenue functions are linear. In reality, total costs are non-linear as it is subject to the law of diminishing returns. The law of diminishing returns states that if one factor of production is increased while the others remain constant, the overall returns will relatively decrease after a certain point.

According to the law of diminishing returns, in a production system with fixed and variable inputs (say, factory size and labor), beyond some point, each additional unit of variable input yields less and less additional output. Conversely, producing one more unit of output costs more and more in variable inputs. The law of diminishing marginal returns states that the production short-run marginal-cost curve will eventually increase.

In reality total cost functions are non-linear and are subject to the law of diminishing returns. This states that as a variable factor (labor) is added to a fixed factor (capital) output will rise and will eventually fall. Figure 2.1 shows that if factor inputs are inefficiently combined, total costs rise rapidly in relation to output. This is because the fixed factor (i.e., capital) is either underutilized or overutilized in relation to the variable factor (i.e., labor). As inputs are more efficiently combined, the rate of increase in total costs slows down as output increases. This shows the influence of the law of diminishing returns on total costs.

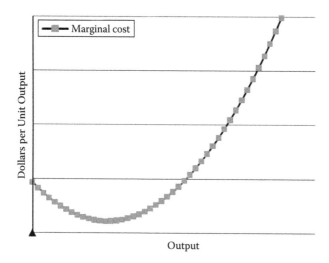

FIGURE 2.1
Widget production: Illustrating law of diminishing returns.

EXAMPLE 2.1

A simple example of diminishing returns in an industrial setting might be a widget factory that features a certain number of square feet of work space and a certain number of machines inside it. Neither the space available nor the number of machines can be added without a long delay for construction or installation, yet it is possible to adjust the amount of labor on short notice by working more shifts or taking on some extra workers per shift. Adding extra man-hours of labor will increase the number of widgets produced, yet only within limits. After a certain point, such things as worker fatigue, increasing difficulties in supervising the large work force, more frequent breakdowns by overutilized machinery, or just plain inefficiency due to overcrowding of the work space begin to take their toll. The marginal returns to each successive increment of labor input get smaller and smaller and ultimately turn negative.

The law of diminishing returns is significant because it is part of the basis that a firm's short-term marginal cost curves will slope upward as the number of units of output increases. And this in turn is an important part of the basis for the law of supply's prediction that the number of units of product that a profit-maximizing firm will wish to sell increases as the price obtainable for that product increases. Other things being held constant, the higher the price of a good (or service), the larger the quantity of that good (or service) that will be offered for sale in a particular time period.

The law of diminishing returns also impacts Lean enterprise including Lean manufacturing and Lean marketing. In reality, it is unlikely that a business can continually sell its output for the same price. Therefore, there comes a point that additional sales can only be increased if the price of the product is lowered. An example is discounts for selling in bulk. Total revenue is also subject to the law of diminishing returns.

2.6   Transfer Function for Average Fixed Cost

Average fixed cost is an important metric for Lean manufacturing. Average fixed cost (AFC) is an economics term to describe the total fixed costs (TFC) divided by the quantity of units produced (Q):

$$AFC = \frac{TFC}{Q} \tag{2.2}$$

Average fixed cost is the total fixed cost per unit of output incurred when a company engages in short-run production. Because average fixed cost is total fixed cost per unit of output, it can be found by dividing total fixed cost by the quantity of output. Average fixed cost is one of three average cost concepts important to short-run production analysis.

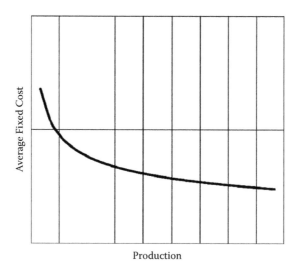

FIGURE 2.2
Average fixed cost decreases with additional production.

Average fixed cost decreases with additional production. The logic behind this relation is relatively simple. Because fixed cost is fixed and does not change with the quantity of output, a given cost is spread more thinly per unit as quantity increases. For example, $1,000 of fixed cost averages out to $10 per unit if only 100 units are produced. If 10,000 units are produced, then the average shrinks to a mere $0.10 per unit.

As shown in Figure 2.2, the average fixed cost curve is negatively sloped. Average fixed cost is relatively high at small quantities of output, then declines as production increases. The more production increases, the more average fixed cost declines. The reason behind this perpetual decline is that a given fixed cost is spread over an increasingly larger quantity of output. This declining average fixed cost curve is a major reason that the average total curve is negatively sloped for relatively small output quantities. In fact, companies that use a lot of fixed inputs relative to variable inputs, such that fixed cost is a substantial share of total cost, spend a lot of their production time in the decreasing portion of the average total cost curve. This has a big impact on how these companies operate. If average total cost declines with additional production, then a company can profitably charge a lower price with increased output.

Average fixed cost, when combined with price, indicates whether a company should shut down production in the short run. If price is greater than average fixed cost, then the firm is able to pay, at least, fixed cost. Even though it might be incurring an economic loss, it will lose less by producing than by shutting down production. If, however, price is less than average fixed cost, then the firm is better off shutting down production.

2.7   Transfer Function for Average Variable Cost

A production system consisting of manufacturing cells linked together with a functionally integrated system for inventory and production control uses less of the key resources needed to make goods. The cost saving from Lean manufacturing often includes reduction of average variable cost.

Average variable cost (AVC) is an economics term to describe the total cost a firm can vary (labor, etc.; total variable coast [TVC]) divided by the total units of output (Q):

$$AVC = \frac{TVC}{Q} \tag{2.3}$$

Average variable cost is the total variable cost per unit of output incurred when a company engages in short-run production. It can be found in two ways. Because average variable cost is total variable cost per unit of output, it can be found by dividing total variable cost by the quantity of output. Alternatively, because total variable cost is the difference between total cost and total fixed cost, average variable cost can be derived by subtracting average fixed cost from average total cost.

In general, average variable cost decreases with additional production at relatively small quantities of output, and then eventually increases with relatively large quantities of output. This pattern is illustrated by a U-shaped average variable cost curve (see Figure 2.3).

The key feature of this average variable cost is the shape. It is U-shaped, meaning it has a negative slope for small quantities of output, reaches a minimum value, then has a positive slope for larger quantities. This U-shape

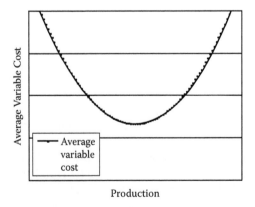

FIGURE 2.3
U-shaped average variable cost curve.

is indirectly attributable to the law of diminishing marginal returns. The U-shape of the average variable cost curve is indirectly caused by increasing, then decreasing marginal returns (and the law of diminishing marginal returns). The negatively sloped portion is attributable to increasing marginal returns and the positively sloped portion is attributable to decreasing marginal returns (and the law of diminishing marginal returns).

Average variable cost, when combined with price, indicates whether a firm should shut down production in the short run. If price is greater than average variable cost, then the firm is able to pay all variable costs and a portion of fixed costs. Even though it might be incurring an economic loss, it will lose less by producing than by shutting down production. If, however, price is less than average variable cost, then the firm is better off shutting down production.

2.8 Transfer Function for Average Total Cost

The Lean manufacturing practitioner does not focus on individual cost factors such as transportation or warehousing, yet rather focuses on "total cost of ownership." For example, with inventory carrying costs representing 15 to 40 percent of total logistics costs for many industries, making decisions based on total cost has dramatic implications for the Lean enterprise.

The average total cost curve is constructed to capture the relation between cost per unit and the level of production. A productively efficient company organizes its factors of production in such a way that the average cost of production is at its lowest point and intersects marginal cost. In the short run, when at least one factor of production is fixed, this occurs at the optimum capacity where it has enjoyed all the possible benefits of specialization and no further opportunities for decreasing costs exist. This is usually not U-shaped; it is a checkmark-shaped curve. This is at the minimum point in the diagram shown in Figure 2.4.

Here, the average cost is equal to total cost divided by the number of goods produced (Q). It is also equal to the sum of average variable costs (total variable costs divided by Q) plus average fixed costs (total fixed costs divided by Q). Average total costs may be dependent on the time period considered (increasing production may be expensive or impossible in the short term, for example). A typical average cost curve will have a U-shape, because fixed costs are all incurred before any production takes place and marginal costs are typically increasing because of diminishing marginal productivity. In this typical case, for low levels of production there are economies of scale: marginal costs are below average costs, so average costs are decreasing as quantity increases. An increasing marginal cost curve will intersect a U-shaped average cost curve at its minimum, after which point the average cost curve begins to slope upward. This is indicative of diseconomies of scale.

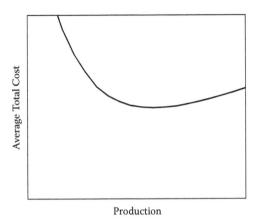

FIGURE 2.4
Average total cost curve.

For further increases in production beyond this minimum, marginal cost is above average costs, so average costs are increasing as quantity increases. An example of this typical case would be a factory designed to produce a specific quantity of widgets per period: below a certain production level, average cost is higher due to underutilized equipment, while above that level, production bottlenecks increase the average cost.

You can combine cost curves to provide information about Firm A's marginal costs that graphically represents the relation between marginal costs incurred by a firm in the short-run product of a good or service and the quantity of output produced. This curve is constructed to capture the relation between marginal cost and the level of output, holding other variables, like technology and resource prices, constant. The marginal cost curve is U-shaped. Marginal cost is relatively high at small quantities of output, then as production increases, declines, reaches a minimum value, then rises. The marginal cost is shown in relation to marginal revenue, the incremental amount of sales that an additional product or service will bring to the firm. This shape of the marginal cost curve is directly attributable to increasing, then decreasing marginal returns (and the law of diminishing marginal returns). If the marginal revenue is above the average total cost price the firm is deriving an economic profit.

As shown in Figure 2.5, we have the following:

- The AFC curve slopes downward and approaches zero on the horizontal axis, while the AVC curve approaches the ATC curve.
- As discussed in Section 2.7, the AFC curve must approach zero, because as production increases, it spreads its fixed costs over a large number of units, so average costs must fall.

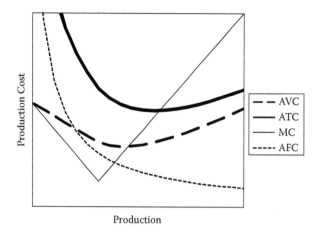

FIGURE 2.5
Transfer function curves for marginal cost (MC), average fixed cost (AFC), average variable cost (AVC), and average total cost (ATC).

- The AVC curve must approach the ATC curves as production increases.
- The MC curve intersects both the AVC and ATC curves at their minimums. If the marginal cost is greater than average total cost, then the average total cost must be rising, and vice versa. Thus, it must be that only when marginal cost equals average total cost that the ATC is at its lowest point. This is a very critical relationship. It means that a company searching for the lowest average total cost of production should look for the level of production at which marginal cost equals average cost.

2.9 Transfer Function for Long-Term Cost Analysis

Thus far we have not considered the long-term production in cost theory. We will now think a bit about the long term, using the concept of average cost. Although the real long-run costs of a product or service are often invisible or hidden in reserve accounts and may never be truly grasped, data analysis using Six Sigma could consistently lead you to better and more rapid solutions.

We have defined the *long term* as "a period long enough so that all inputs are variable." This includes, in particular, capital, plant, equipment, and other investments that represent long-term commitments. Thus, here is another

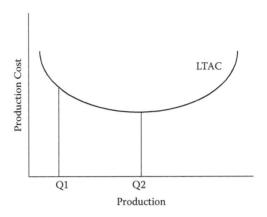

FIGURE 2.6
Transfer function curves for long-term cost analysis (LTAC).

way to think of the long run: it is the perspective of product development and production expansion planning.

So let's approach it this way: Suppose you were planning to build a new plant, perhaps to set up a whole new company, and you know about how much output you will be producing. Then you want to build your plant so as to produce that amount at the lowest possible average cost. The long-term average cost, LTAC, curve of a firm shows the minimum or lowest average total cost at which a firm can produce any given level of output in the long run (when all inputs are variable).

As shown in Figure 2.6, the long-run average cost curve depicts the per unit cost of producing a good or service in the long run when all inputs are variable. The curve is created as an envelope of an infinite number of short-run average total cost curves. The LTAC curve is U-shaped, reflecting economies of scale when negatively sloped and diseconomies of scale when positively sloped.

The reason for the U-shape of the long-term average cost curve is not the law of diminishing returns. Instead, the explanation lies in understanding one of the most important concepts in production known as economies of production scale. Economies of production scale are said to exist when the per-unit production cost of all inputs decreases as production increases. As to why such economies of production scale may exist, they may be traced to such factors as increased labor and managerial specialization and more efficient capital use. Finally, economies of production scale are not necessarily present in all industries.

In the long term, when all factors of production can be changed, the scale of the enterprise can be increased. In this case productive efficiency occurs at the optimum scale of production where all the possible economies of scale have been enjoyed and the firm is not large enough to experience

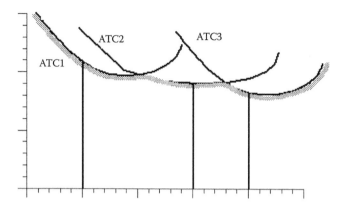

FIGURE 2.7
Lower envelope curve for long-term cost analysis (LTAC).

diseconomies of scale. Here, the long-term average cost curve is the envelope of the short-term average cost curves. For example, for any given plant scale, capital inputs are fixed in the short term and there is a point on the average total cost curves where average total cost is minimized. Now if you build a bigger plant, production will increase, and there will be another short-term ATC curve created. And each point on this bumpy planning curve shows the least unit cost obtainable for any production when the company has had time to make all desired changes in plant size.

To make it a little simpler we will suppose that you have to pick just one of three plant sizes: small, medium, and large. In Figure 2.7 are the average cost curves for the small (ATC1), medium (ATC2), and large (ATC3) plant sizes. Also, as shown in Figure 2.7, we have the following:

- If you produce a small number of units, the small plant size gives the lowest cost.
- If you produce a medium number of units, the medium plant size gives the lowest cost.
- If you produce a large number of units, the large plant size gives you the lowest cost.

Therefore, the LTAC—the lowest average cost for each output range—is described by the lower envelope curve, shown by Figure 2.7's thick, shaded curve that follows the lowest of the three short-run curves in each range.

Six Sigma is a statistical and data-driven process for eliminating defects and to ensure higher (and sustainable) productivity and efficiency in every line of a business over the long term. If time is spent up front to do it right, time can be saved in the long term, quality increases, there are fewer defects, and the LTAC is significantly reduced. Six Sigma helps companies bridge the gap between strategy and operations by providing predictive, in-process performance measures (leading indicators) that can be linked to business goals and outcomes (lagging indicators). Also, some companies use Six Sigma to link their business and unit-level scorecards, including short-term and long-term cost savings, to their corporate goals.

2.10   Goal Tree and Success Analysis: From Goal Setting to Successful Goal Setting

Goals have been widely and successfully used in requirements engineering. A goal can be defined as an objective or state that must be reached, and its definition makes reference to a set of properties whose fulfillment must be guaranteed. When developing a Lean manufacturing system, a goal tree and success tree analysis takes the business environment into account so that the system properly fulfils the needs of the organization. This approach is based on the mapping of business process goals into system goals from which requirements are defined. A goal tree and success tree analysis for a Lean manufacturing system is performed as follows:

- The goal tree for a Lean manufacturing system is constructed by decomposing the overall goal of the Lean manufacturing system into a set of necessary and sufficient subgoals, and continuing the task of decomposition for each subgoal until physical components are needed to satisfy the subgoal (see Figure 2.8).
- At this point a success tree for that particular subgoal begins. In order to construct the success tree, all different paths by which the subgoal is attained must be represented. The difference between the success tree and the goal tree is that only one of the success paths must be satisfied for a subgoal to be achieved (see Figure 2.9).

Manufacturing processes have goals that must be fulfilled during or after their execution. There are subgoals that denote important milestones within the process and whose fulfillment is possible due to the actions of all the participants involved. These subgoals are operational goals that indicate when a process instance can be considered completed.

As shown in Figure 2.8 and Figure 2.9, each block of the goal tree represents a goal that can have conditions and attributes associated with it. The conditions create a dynamic goal tree that changes in accordance with the new conditions for Lean manufacturing. The attributes allow further description of the goal such as its priority or the order (relative to the others of its level) in which it must be satisfied. Conflicts among and order of the goals are hence expressed by attaching conditions and attributes to the goal.

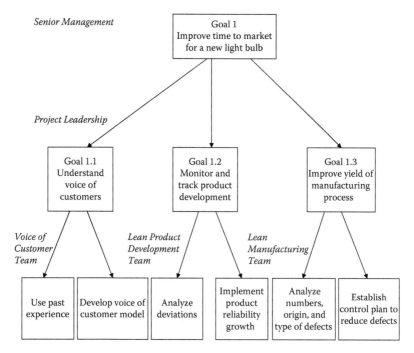

FIGURE 2.8
A goal tree for developing a new light bulb.

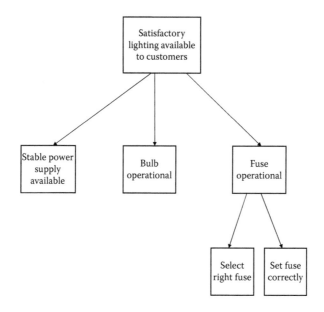

FIGURE 2.9
A goal tree for assuring new light bulb reliability.

GOAL TREE EXERCISE

This exercise describes an activity that has its roots (literally) in industrial psychology. The goal is to look at what goals the participants have and to identify what steps need to be taken to make these goals realities through the language of metaphor. All through the exercise, it is encouraged for group members to talk among themselves, to the facilitator(s), or to the group at-large in order to gain a more complete understanding of their goals and the influences upon them.

PROCESS

What is needed for this exercise is paper and writing or drawing tools such as markers or crayons. The first step is to identify a goal. Make sure that goals are realistic and measurable. Have the participants write down on the top or bottom of the paper what their goals are. Next, we are going to look at the parts of the goal tree and how each of them affects the goal.

OUTSIDE INFLUENCES

A growing tree has many influences that affect its growth from the world around it. Rain and sun help to sustain the tree, and wind serves to shape it somewhat. Fire and severe wind can serve to destroy the tree. Have the participants think about what influences, good or bad, are affecting their goals. Identify them and draw them onto the paper. They might want to have sun and rain reflect the good influences. Bad influences can be seen as fire or pollution.

TRUNK

The trunk of the tree represents the steps we need to achieve to make the goals manifest. Have the participants reflect upon their goals, and break them into manageable baby steps. This is an ideal opportunity for group brainstorming.

FRUIT

Next, have participants draw the leafy part of the tree and pieces of fruit growing in the tree. The fruit represents the goal achieved. Have the participants label the pieces of fruit to represent signs that the goal has been achieved. One might reflect money, another leisure time, or whatever they might see as a manifestation of their goal (thus the need for goals to be measurable).

PRESENTATION

The final step would be to have the participants report to the big group what their individual goal plans are by showing their tree and talking about what the parts represent.

In an organizational scope where manufacturing is based on process analysis and improvement, development methodologies for manufacturing process support systems differ from traditional ones. Therefore, organizational concerns must be taken into account and requirements engineering approaches must provide new ways of elicitation. The design of a Lean manufacturing system is a complex process. A good understanding of design for production issues is essential to ensure a quality product at optimal costs. The overall objective is to improve the design process so that performance and producibility requirements can be met simultaneously. The following chapters will discuss the big picture Lean manufacturing technique to consider the design and production process in a systematic manner and thus to feed back producibility requirements into the initial design phase. The adopted approach explicitly questions linkages between product attributes such as cost or performance, and the ability of the design and production specification to ensure delivery of the attributes.

3

Understanding the Voice of Customers: The Essential Elements

This chapter enables you to

- Understand the transfer function of production cost and analyze how companies produce and offer goods for sale
- Recognize the difference between short-term and long-term cost functions, marginal cost function, and the shapes of various cost curves
- Analyze industrial sectors to prevent production failures and win in the market

3.1 Voice of Customers: Relentlessly Customer Focused

This chapter describes the need for manufacturers to become more customer focused and summarizes the benefits of better understanding the voice of customers. The voice of customers (VOC) model puts the customer, not the producer, first in a quest to deliver world-class manufactured products.

In an increasingly competitive global market, customer expectations have risen and providers across all sectors need to step up to the mark. For manufacturers, there is also a need to design products that meet the needs of an increasingly diverse customer base. The challenge now is how to achieve this.

It is important for manufacturers to get this right. If a company does not understand what really matters to the customers, the company will waste time and money. The company will also risk compromising its reputation by offering products that customers do not recognize as being relevant to them and have difficulty accessing. If a company does make the effort to engage systematically in understanding the needs and behaviors of customers, the rewards will be felt economically and socially.

Customer engagement becomes a reality with the following business benefits:

- The customer voice can be fed back to the production management and acted on.

- Frontline workers can feel more empowered because their views are heard and reflected in product design and delivery.
- Product designers can be more confident about getting the right outcome in the best way for all concerned.

> The voice of customers can be defined as a deep "truth" about the customer based on their behavior, experiences, beliefs, needs or desires, that is relevant to the task or issue and "rings bells" with target people.

Besides the definition, manufacturers also need to consider implementing VOC as a discipline. Here, VOC capability is defined as

> Having a deep, embedded knowledge of the customers and the market that helps structure thinking and sound decision making for manufacturing.

The discipline comes from a combination of multiple pieces of data, built into a joined-up "big picture" through bottom line–based business analysis.

Listening to the VOC is not just a matter of implementing Lean manufacturing toolboxes or finding more effective ways to measure and increase customer satisfaction levels. Tools and technology are important. But they are not enough. That's because listening to the VOC is a journey that the whole organization needs to make. Listening to the VOC consistently embraces the following three concepts:

- First, companies know they can become customer focused only if they learn everything about their customers at the most granular level, creating a comprehensive picture of each customer's needs—past, present, and future.
- Second, companies know that this picture is useless if employees cannot or will not share what they learn about customers, either because it is inconvenient or because it does not serve their interests.
- Finally, companies use this insight to guide not only design and production decisions but also manufacturing strategy and organizational structure as well.

Clearly, employees need to work individually and collectively to help their company at a central and local level to establish a customer-focused culture of continuous improvement.

VOC is not a new discipline. However, in recent years we have witnessed a clear cultural shift in the service sector—led by brand-leading financial services, telecommunications, and fast moving consumer goods companies—where customer insight as a discipline has moved into the heart of companies and their operations. This has resulted in part from the pressure of competition to deliver the right services in environments where customers are more demanding and the pressure on costs is constantly increasing. The result has been the development of sophisticated insight capabilities within organizations where innovative qualitative techniques are routinely applied alongside tried and tested quantitative research to generate customer stories that have impact at the highest levels.

Similarly some manufacturing sectors have enthusiastically embraced the principles and are applying VOC in product design and production. However, in many organizations, VOC is still misunderstood. Customer intelligence exists within manufacturing organizations, sometimes in large volumes, from surveys and consultations, customer databases, Web usage data, and the front line (among other sources). Yet, often absent is the coordinating function that links the information together to generate stories and a genuine VOC that meet specific strategic and business objectives, and ensure that these are understood and embedded in manufacturing processes.

This chapter sets out the basic elements required to establish an effective VOC capability. These elements will need to be adapted to meet the needs of specific manufacturing organizations. VOC is a business activity more than a research activity. Although based on robust techniques, it is less concerned with methodological purity. It should not be limited to a single team or department but should span an organization, empowering staff to make confident decisions affecting strategy, policy, and delivery that are grounded in VOC and managed risk.

After studying the following sections, you should understand what might be involved in setting up an effective VOC capability from a process and structural perspective, as well as the type of activity for which a manufacturing function should take responsibility.

3.2 Voice-of-Customers Capability: Critical Elements

This chapter sets out the critical factors for establishing an effective VOC capability. An effective VOC capability:

- Has top-level leadership and sponsorship
- Draws on customer information from multiple sources and turns it into stories that have business value
- Is independent but with traction across the organization
- Values customer insight as a strategic asset informing policy, strategy, operations, and communications
- Involves the right level and mix of skills and experience

3.2.1   Top-Level Leadership and Sponsorship

Support at the executive level management is very important. Executive level management sets the vision for where the organization needs to be and determines the needs of customers and clients that the organization should strive to meet. The leadership of any manufacturing organization therefore needs to understand, value, and champion the role that VOC can play in achieving positive outcomes for customers and clients.

Once accountability for customers is in place at the executive level, executives will need support from a group or network drawing on members from across the organization (including policy, strategy, communications, and operations). This group is responsible for the following:

- Agreeing on the VOC work plan
- Setting priorities in line with the organization's overall business strategy
- Overseeing progress on major VOC projects
- Taking action based on the output of VOC work
- Ensuring results and impacts are fed back to their teams
- Drawing on external views from customers and from VOC teams in other organizations

Finally, top-level leadership will be responsible for creating the all-important customer-focused *culture,* where employees understand who their customers are, routinely share information and stories about them, and make decisions that are rooted in a deep customer understanding.

3.2.2   Turn Voice of Customers into Stories with Business Values

VOC draws on multiple sources of customer information to generate meaningful stories. These stories have business value and can be acted on. This is the value that a VOC function can add. A VOC "function" acts as one recognized place where VOC is made available, offering a strategic overview of the various sources of customer information available to an organization. Customer information often exists in large volumes, from surveys to focused-group research, consultation data, and customer complaints. The VOC function is about pulling together the information as required from many sources, not necessarily owning them.

The VOC team should optimize existing resources by collating and analyzing the information that already exists. Many companies have a number of teams that deliver different parts of the jigsaw puzzle. To deliver effective VOC and to see the whole picture, the VOC team will need access to all of these. Further, the VOC team has a responsibility to ensure that all existing sources of VOC are properly exploited. For example, this means gathering, analyzing, and applying lessons from the valuable information contained within the hundreds of letters of complaints received annually at warranty and service departments; or learning from existing research before commissioning new studies.

It is also important that the VOC team not only gathers the intelligence already available but also shapes the information that the organization routinely collects. Essentially, this is about ensuring that the organization is asking the right questions about the things that really matter to customers and checking that the right things are being measured.

3.2.3 Independent but with Traction Across the Organization and Beyond

The VOC team has an independent customer advocacy role, providing a source of challenge for the rest of the organization from the customer perspective. Regardless of where it is situated within the organizational structure, it is the role of the VOC function to act as the impartial VOC across the business—even if it means delivering unpalatable messages, which may conflict with received wisdom. This means that the team may often work horizontally, delivering customer insight across the organization, not in service of, or prioritizing, any specific unit.

It is also important that the insight function be outward looking, drawing on information about the customer and published information from other organizations and international experience. The aim here should be to pick up on interesting new ideas, to draw parallels and most critically to spot opportunities for synergy and for teaming up to improve customer experiences across the board. The VOC team should never take a purely manufacturing-centric view but should look at the system or journey in the way that the customer experiences it. Packaging up insight to focus only on parts of the system that a single department delivers simply won't drive the necessary improvements from a customer perspective.

Finally, the VOC function acts as a means by which VOC can be stored and shared within and between organizations in service of a range of different but linked objectives. VOCs that have been sourced in service of one manufacturing area, for example, may be equally applicable to others. Organizations will achieve cost savings by exploiting the full potential of existing sources before commissioning new research.

3.2.4 Strategic Asset Informing Policy, Strategy, Operations, and Communications

Effective VOC is informed by, and should inform, strategy, policy, operations, and communications priority areas to ensure relevance and correct ownership. If the link between the VOC and the manufacturing business objectives is unclear, don't do it. Recognizing VOC as a strategic asset and incorporating it into production planning as an integral part of the manufacturing business process—whether in policy, operations, or communications—are vital.

Effective VOC should help to inform what is appropriate to measure. Generating VOC is essentially a business process, aimed at creating something that has value to the organization. Once VOC has helped to formulate those measures, meaningful targets can be set. The policy and strategy development frameworks themselves should include links at every stage to the appropriate VOC work streams, showing where VOC has been considered and incorporated into findings and recommendations. In product development processes this means sourcing from the customer at every stage, including concept, "prototype," and usability testing.

3.2.5   Right Level and Mix of Skills and Experience in the Team and Wider Networks

VOC is a discipline that requires specialist as well as generalist skills. The team should include a mix of professional backgrounds and experience to encourage innovation and stretch thinking, for example:

- Consultancy skills to clarify—and challenge if needed—the manufacturing business objectives that drive VOC work and to feed back results, ensuring that they are understood and can be acted upon.
- Strong communication and networking skills to work effectively with teams in strategy, policy, operations, human resources, communications, and project areas.
- Direct links to the front line to be as close to the customer as possible.
- Excellent marketing and strategy capability (understanding VOC in different forms, use VOC tools, apply in production planning process, etc.). Team members need not be practitioners of all VOC tools, yet they should have good analytical skills so that they can understand their application, know when to bring in outside help, and be able to specify objectives. The team also needs to understand the language of specialists to be able to turn data into insight.
- An ability to engage with business teams, to understand how they work, and to talk the same language in order to position VOC in the appropriate context for each individual circumstance.
- Be well networked within and outside the organization, bringing in ideas from other departments and organizations in the public, private, and voluntary sectors, as appropriate.
- Make the right linkages and be able to draw on the right skills elsewhere in the organization, for example, from sales, statisticians, and researchers. Not all of the skills need to sit in the immediate team, yet the right links and ways of working do need to be in place.

3.3 Voice of Customers' Role and Responsibilities

The core activities and responsibilities for an effective VOC capability are summarized as follows:

- Reporting at a senior (board or director) level to advise on and help to develop manufacturing business priorities and strategy
- Advising on the appropriate range of VOC tools and working with relevant parties to develop clear objectives that translates into VOC objectives
- Helping to define what action is going to be taken as a result and being clear that the output of VOC activity is the starting point, not the ending point
- Investigating the links between VOC and manufacturing outcomes, and ensuring that these are established, measured, and tracked
- Ensuring all VOC (including research and evaluation projects) fits and informs the organization's manufacturing business strategy, as well as the overall VOC program
- Acting as a formal gatekeeper (with power of veto) for major VOC work including research and evaluation programs
- Providing consultancy support, advice, and expertise for all research, evaluation, and VOC activity, ensuring value for money
- Evaluating existing formal training in VOC within the organization and providing training support to embed an understanding across the organization of what VOC means and how it should be applied
- Coordinating the gathering of VOC across key project areas on an organization-wide level, with close alignment to production planning, research, and communications in particular
- Disseminating key findings of VOC and research programmes across the organization in a timely way, ensuring that the information is widely understood and actionable; this may mean presenting the information and key messages in a range of forms and across a range of channels
- Communicating VOC and research findings outside the manufacturing organization, where appropriate, to partners and related organizations, and collaborating with these organizations where applicable on joint insight work
- Providing a major repository for VOC, research, and evaluation from across the organization and from relevant external sources, so that research information is stored in one place and can be easily accessed and disseminated

Developing a customer insight capability is likely to be an iterative process for manufacturing organizations. The recent introduction of VOC functions in manufacturers demonstrates that there is momentum building around this work and that existing VOC resources are proving their value. Now is the time to raise our sights and also to ensure that manufacturers' VOC capabilities are more than just the sum of their parts.

3.4   Analyzing Voice of Customers: Law of Demand

Understanding the behavioral responses of individuals to the actions of business is an ongoing need of companies providing manufactured products within a competitive environment. Choosing one product over another, choosing to buy or not to buy is an active response on the part of a consumer. The development of a formal structure from which to explore these choices holds the promise of accurately anticipating the likely consumer response to a given product without having to actually release the new or modified product for production. The ability to predict uptake and usage of a product enables new or modified product that will lead to optimum bottom-line revenue.

Such accurate predictions are critical for manufacturing businesses. Being able to determine the elements of consumer demand provides a formidable advantage to any company or organization seeking a competitive advantage. This in turn provides the opportunity to create products with strong appeal and to understand the value appeal of each considered product attribute. The consumer theory is to explain the consumption behavior of consumers. Starting from the postulates, economists build a process of logical deduction to form the theory of consumers so as to deduce and explain the so-called law of demand.

In reality consumers are faced with various kinds of goods under its subjective level of preference and choice. At the same time, they could only be satisfied by the presence of enough money or the effective demand, that is, sufficient purchasing power. Consumers are trying to maximize their utility, where utility is satisfaction. Most consumers are constrained by a limited budget. Thus, utility maximization is a constrained optimization problem. With an aim of "utility" maximization, the consumer will be in an optimum state if and only if the purchasing power of money can effectively bring the consumer to a higher ranking of preference (a state of higher and higher level of satisfaction) until all income is used up.

For example, suppose you are hosting a party at Gianni's Pizza. You have a limited party budget and you have to decide how to allocate this budget between pizza, beer, and salad. Your objective is to maximize the utility your guests receive from the food and libations at the party. We could determine the total utility (TU) guests receive from different quantities of pizza, beer, and salad. Then we could calculate the marginal utility (MU) from one more pizza, one more beer, and one more salad. Finally, we could divide these marginal utilities by the price of the respective good to get the marginal utility per dollar spent (for the last dollar spent on each good). From here the consumer behavior is viewed from two aspects:

1. The subjective choice and level of preference are condensed under the concept of indifference curve in the ordinal utility theory.
2. The objective purchasing power is revealed by the budget constraint.

These two aspects together give the equilibrium and optimum state of the consumer—the state of utility maximization. From the optimal state of the consumer, the analysis can go further to seek the relation between price and quantity demanded of a consumer within a given price range. As a result, the law of demand can be examined in detail.

3.4.1 The Law of Diminishing Marginal Utility

What trend would you expect to see in the marginal utility per dollar spent as we obtain more of each good? The totals might increase, yet they would increase at a decreasing rate. This is called diminishing marginal utility and we expect it due to satiation. We get more enjoyment out of the first bite of pizza than out of the hundredth bite, and after a while additional bites make us worse off (too full or sick). This is a phenomenon typically observed in consumption.

Based on the law of diminishing marginal utility, VOC is the satisfaction of customers from consumption that can be measurable (i.e., be quantified) and discernible (i.e., comparable). From the observation of real-life situations, the theory suggests that the total utility of a consumer will increase through consumption, but for successive units of the goods consumed (the additional or extra units of utility received) the marginal utility will gradually diminish. When anyone uses the term *marginal utility* it already implies that utility is assumed to be measurable. Otherwise the concept of marginality cannot be applied.

3.4.2 The Law of Equi-Marginal Utility per Dollar

Inaccurate estimate or sensing of the system state can lead to incorrect decisions, with consequent adverse effects on manufacturing performance. It suggests that when a consumer buys more of a good, its marginal utility on the good decreases, but at the same time, other goods will be consumed less if income is fixed. The rationale is that as long as the marginal utility of any two or more goods is different, a consumer will try to consume the good with a higher marginal utility.

Example 3.1

Given: income = $14; price of A = $1; price of B = $2.

Quantity (Q)	Good A		Good B	
	T U	*M U*	*T U*	*M U*
0	0	—	0	—
1	10	10	24	24
2	19	9	45	21
3	27	8	64	19
4	34	7	80	16
5	40	6	94	14
6	45	5	104	10
7	49	4	110	6
8	52	3	114	4

Possibilities		
Q_A	Q_B	*T U* of A and B
0	7	0 + 110 = 110
2	6	19 + 104 = 123
4	5	34 + 94 = 128
6	4	45 + 80 = 125
8	3	52 + 64 = 116

The result is

MU of A/Price of A = MU of B/Price of B = 7 units of utility/$1

Once we have constructed the table of marginal utilities per dollar, we could find the combination of pizza, beer, and salad that maximizes our utility subject to our budget constraint. What would be the relationship between the three goods at the optimal point?

$$MUp/Pp = MUb/Pb = MUs/Ps$$

where
  MUp = Marginal utility of pizza
    Pp = Price of pizza
  MUb = Marginal utility of beer
    Pb = Price of beer
  MUs = Marginal utility of salad
    Ps = Price of salad

This makes intuitive sense. We will always buy more of the good that gives us the greatest relative bargain in utility. Thus, we will keep readjusting our combination if the marginal utility per dollar for the last dollar spent is greater for one good than it is for all the others. Because of diminishing marginal utility, as we purchase more of the good where MU/P is greatest, MU (and hence MU/P) decreases. Similarly, MU (and hence MU/P) increases for the goods we buy less of. Thus, shifting expenditures to the good with the highest relative MU/P serves to equalize MU/P across all goods.

3.4.3 Condition of Consumer Optimum: Utility Maximization

From Example 3.1, the consumer will consume a different quantity of good A and B. The MU (obtained by the last dollar spent) derived from the good A and B will be equal so that a state of equilibrium could be reached.

$$MU_X/P_X$$

$$= MU_Y/P_Y$$

$$= MU_Z/P_Z \text{ (state of consumer optimum)}$$

If the equation is rewritten into another form:

$$MU_X/MU_Y$$

$$= P_X/P_Y$$

(The ratio of MU of any two goods equals their **relative** price.)

3.5 Maximizing Utility to Customers

The ordinal theory suggests that utility is only relatively discernible but not quantifiable. Utility is, in fact, a series of assigned numbers to rank options by the consumer preference. The assigned numbers reveal what is more preferred but cannot tell how much the difference is. In other words, utility can only be ranked by an order or a scale of preference to show the degree of willingness of a consumer. Consumer preference denotes an observation pattern of choice, whereas utility is an ordinal scale constructed to represent that regular pattern. From here, it becomes the axioms or propositions on the assumption of rationality:

1. The consumer is capable of comparison and makes substitution on goods to show his indifference on the goods consumed.
2. The consumer must have a scale of preference in mind before he purchases. He is consistent in buying and also clear about his different level of satisfaction (but he cannot tell how much). Satisfaction

can be obtained through the consumption of different goods, that is, there is the possibility of transitivity.

3. Utility maximization and a state of optimum are revealed by the fact that the consumer always prefers more to less.

3.5.1 Use Indifference Curve to Represent Customers' Level of Preference

Based on these assertions, F. Y. Edgeworth (1845–1926) first suggested the indifference curve to represent the level of preference (or satisfaction) a consumer has when two goods are consumed in different amounts, but each combination of these two goods yields the same level of preference. The properties of the indifference curve include:

1. It is the locus of the combination of two goods that are equally satisfied to a consumer or to which the consumer is indifferent.

2. The slope of this curve is negative; there is some degree of substitution between the two goods.

3. The curve is convex to the origin, that is, the marginal rate of substitution (MRS) of two goods is diminishing. MRS in consumption (= DY/DX) equals the number of a good Y that had to be given up for each unit of good X to maintain the same level of utility along any point on the indifferent curve.

4. A curve farther away from the origin means that it stands for a higher level of preference than the one nearer the origin. Again, the magnitude between any two indifference curves does not matter.

5. As a consumer changes his choice in a continuous process, there must be at least another curve between any two indifference curves, that is, there may be an infinite number of curves for a single consumer on a good.

6. There is no intersection for any two curves in the indifference map. It is therefore only useful to compare points on the same curve. The marginal rate of substitution in consumption is a measure of change along the curve only, not the shift of the curve because different curves represent different levels of preference and cannot be compared.

The different shapes of the indifference curves indicate different degrees of substitution of the good. To a consumer, goods can be closed substituted, completely substituted, or simply no substitution at all.

3.5.1.1 Degree of Substitution

Ordinary goods have a normal indifference curve that slopes downward. As shown by Figure 3.1, the slope of an indifference curve (in absolute value), known by economists as the marginal rate of substitution, shows

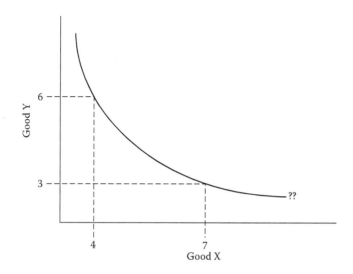

FIGURE 3.1
Example of an indifference curve.

the rate at which consumers are willing to give up one good in exchange for more of the other good. For most goods the marginal rate of substitution is not constant so their indifference curves are curved. The curves are convex to the origin, describing the negative substitution effect. As price rises for a fixed income, the consumer seeks less the expensive substitute at a lower indifference curve. The substitution effect is reinforced through the income effect of lower real income (Beattie–LaFrance). An example of a utility function that generates indifference curves of this kind is the Cobb–Douglas function

$$U(x, y) = x^a y^{(1-a)}$$ (3.1)

3.5.1.2 Perfect Substitution

Let's say you are willing to trade one dime for two nickels, therefore your marginal rate of substitution between nickels and dimes would be a fixed number, that is, 2. Because the MRS is constant, the indifference curves are straight lines. If the goods are perfect substitutes then the indifference curves will be parallel lines since the consumer would be willing to trade at a fixed ratio (see Equation 3.2). The marginal rate of substitution is constant.

An example of a utility function that is associated with indifference curves like these would be

$$U(x, y) = ax + ay$$ (3.2)

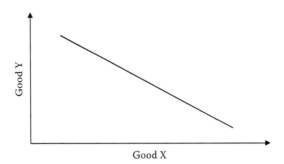

FIGURE 3.2
An indifference curve where goods X and Y are perfect substitutes.

3.5.1.3   *Perfect Complements*

An example of perfect complements is a left shoe and a right shoe. Suppose now that someone offered you bundles of shoes. Some of the shoes fit your left foot, others your right foot. How would you rank these different bundles?

- You would judge a bundle based on the number of pairs you could assemble from it. A bundle of five left shoes and seven right shoes yields only five pairs. Getting one more right shoe has no value if there is no left shoe to go with it.
- Goods that are perfect complements have L-shaped indifference curves (see Figure 3.3). The consumer is no better off having several right shoes if she has only one left shoe. Additional right shoes have zero marginal utility without more left shoes. The marginal rate of substitution is either zero or infinite.

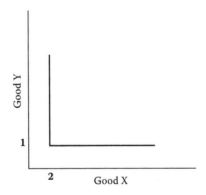

FIGURE 3.3
Indifference curves for perfect complements X and Y.

Another example of perfect complements would be if you had a cookie recipe that called for three cups flour to one cup sugar. No matter how much extra flour you had, you still could not make more cookie dough without more sugar. An example of the type of utility function that has an indifference map like that above is:

$$U(x, y) = \min(\alpha x, \beta y) \tag{3.3}$$

3.5.2 Use Budget Line to Reflect Customers' Purchase Power

The objective purchasing power in the form of income is represented by the budget line between two goods. The slope of the line gives the relative price of any one good. It tells what the consumer must give up in terms of another good in order to buy one good. The slope of the budget line is called the marginal rate of substitution in exchange (P_X/P_Y).

The concept of relative price is important because a rise in relative price would encourage the producer to put more resources in production. The concept also conveys the market information of relative scarcity of those resources. The budget line rotates when the relative price changes. The shift of the line means that either the income changes or there is a change in the price of both goods. Both cases also imply a change in the purchasing power of the consumer.

3.5.2.1 Budget Line, Consumption Possibility Line, Budget Constraint

- The budget line plots combinations of goods that spend all the individual's income. Essentially, the budget line is a menu showing what the consumer can afford.
- Divisible goods can be purchased in any quantity desired (gasoline, for example); indivisible goods must be bought in integer units (services, for instance). The model we develop is based on the assumption that all goods and services are divisible.
- The consumption possibilities are all the combinations on a budget line.
- Points above and to the right of the budget line are unobtainable/ unaffordable.
- Any points on the line and inside the budget line are obtainable/ affordable.
- The slope of the budget line depends upon the relative prices of the two products.

EXAMPLE 3.2

Lisa has a budget of RM30 a month to spend. She buys two goods: pizzas (cost RM6 each) and soda (RM3 for a bottle). If Lisa spends all of her income, she will reach the limits to her consumption of pizzas and soda as follows:

Consumption Possibility	RM6 Pizzas (per Month)	RM3 Soda (Bottles per Month)
A	0	10
B	1	8
C	2	6
D	3	4
E	4	2
F	5	0

3.5.2.2 Budget Equation

The budget equation is a mathematical formula showing affordable combinations of goods and services. For two goods, soda and movies, the budget equation is

$$Income = Expenditure$$

$$Y = (Psoda \times Qsoda) + (Ppizza \times Qpizza)$$

$$Q_{soda} = \frac{Y}{P_{soda}} \quad \frac{P_{pizza}}{P_{soda}} (Q_{pizza})$$

where
    $Qsoda$ = Quantity of sodas purchased
        Y = Consumer's income
   $Psoda$ = Price of a soda
  $Ppizza$ = Price of a movie
  $Qpizza$ = Quantity of movie purchased

The vertical intercept in the budget line ($Y/Psoda$) is the consumer's real income in terms of sodas. A household's real income is the maximum quantity of a good that the household can afford to buy. The magnitude of the slope of the budget line ($Ppizza/Psoda$) is the relative price of a pizza in terms of a soda. This relative price shows how many sodas must be sacrificed to eat an additional pizza.

3.5.2.3 Shifts of Budget Line

The budget line shifts if the price of a good or if income changes.

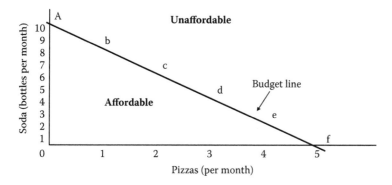

FIGURE 3.4
The budget line.

3.5.2.4 A Change in Prices

A drop in the price of a good on the horizontal axis (pizzas) causes the budget equation to rotate out, around the (unchanging) vertical intercept. The slope of the budget line becomes flatter.

An increase in the price of a good on the horizontal axis (pizzas) causes the budget line to rotate inward and become steeper.

Referring to Example 3.2, if the price of a pizza falls from RM6 to RM3, real income in terms of soda does not change but the relative price of a pizza falls. The budget line rotates outward and becomes flatter. If the price of a pizza rises from RM6 to RM12, the relative price of a pizza increases and the budget line rotates inward and becomes steeper as shown in Figure 3.5.

3.5.2.5 A Change in Income

An increase in the consumer's money income shifts the budget line rightward. The slope of the budget line does not change (parallel), because a change in money income changes real income but does not change relative price.

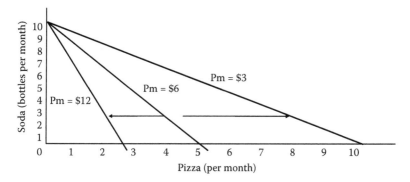

FIGURE 3.5
Lisa's budget lines (a change in price).

The bigger a household's money income, the bigger is real income and the farther to the right is. The smaller a household's money income, the smaller is real income and the budget line is closer to the origin than the initial.

Referring to Example 3.2, the following new budget lines show how much Lisa can consume (while the prices of pizza and soda remain constant) if her income changes to (a) RM15 a month and (b) RM60 a month.

If Lisa's income falls to RM15 a month, the new budget line is parallel (has the same slope, the relative price is the same) to the initial one but closer to the origin, because her real income has decreased.

3.5.2.6 Consumer Optimum

The indifference curve and the budget line together constitute the consumption behavior. Graphically speaking, the two curves meet at a point where the indifference curve is tangent by the budget line to get a *unique* or *internal* solution. This point of tangency represents the highest level of preference obtained by a person given a fixed amount of income. This point is also the point of optimum condition or *utility maximization*.

In mathematics, the slopes of the indifference curve and the budget line are the same.

$$\text{Slope of the budget line} = \text{MRS in exchange} = P_X/P_Y$$

$$\text{Slope of the indifference curve} = \text{MRS in consumption} = \Delta Y/\Delta X$$

In equilibrium,

$$P_X/P_Y = \Delta Y/\Delta X$$

Consumers typically diversify in consumption. They usually purchase a basket of goods and services. Only with the use of convex indifference curves

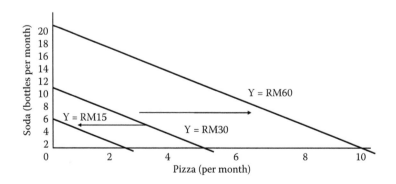

FIGURE 3.6
Lisa's budget lines (a change in income).

can we obtain an *interior* solution in the analysis of consumer optimization. It gives results relevant to the real-world phenomena.

3.6 Kano Model: Integrate the Elements for Voice of Customers

Integrating the elements for VOC, Noritaki Kano developed the model of the relationship between customer satisfaction and quality. The Kano model of customer satisfaction can explain the VOC. As shown by Figure 3.7, the Kano model shows that there is a "must-be" level of quality that customers assume that the product will have and can be summarized as follows:

- Must-be attributes—The must-be attribute curve lies in the lower half of the chart, representing a feeling of dissatisfaction in the customers. These attributes are basic criteria of a product. If the product or service doesn't meet the need sufficiently, the customers become very dissatisfied. On the other hand, as the customer takes these requirements for granted, their fulfillment will not increase her or his satisfaction. Fulfilling the must-be attributes will only lead to a state of not dissatisfied. The customer regards the must-be attributes as prerequisites; she or he takes them for granted and therefore does not explicitly demand them. Must-be requirements are in any case a decisive competitive factor, and if they are not fulfilled, the customer will not be interested in the product at all.

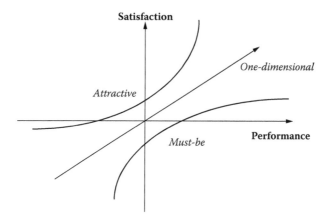

FIGURE 3.7
The Kano model.

- One-dimensional "expected" attributes—The one-dimensional expected attributes line graphs expectations that are actually considered by customers as a measure of quality of service provided. The customers are thus not satisfied if their expectations are not met. With regard to one-dimensional attributes, customer satisfaction is proportional to the level of fulfillment. It means that it results in customer satisfaction when fulfilled and dissatisfaction when not fulfilled. The level of customer satisfaction increases linearly with the level of fulfillment. The customer usually explicitly demands these attributes. For example, when customers want to buy a new car, "mileage" may be such an attribute.
- Attractive attributes—Attractive attributes are something more than expected, which can delight customers. These attributes are the product criteria, which have the greatest influence on how satisfied a customer will be with a given product. These attributes are neither explicitly expressed nor expected by the customer. Although fulfilling these requirements leads to more than proportional satisfaction, their absence does not cause dissatisfaction because customers are unaware of what they are missing.

Simply tracking competitor progress is not enough. We also need to assess customer satisfaction levels. Six Sigma addresses only the portion of the Kano model on and below the line labeled "One-Dimensional Expected Attributes." Long-term success for an organization requires that the company innovate in product and production using Lean manufacturing, which we will further explore in the next chapter.

4

Balance Production and Demand: Value Stream Mapping

This chapter enables you to

- Understand the transfer function of production cost and analyze how companies produce and offer goods for sales
- Recognize the difference between short-term and long-term cost functions, marginal cost function, and the shapes of various cost curves
- Analyze your industrial sectors to prevent production failures and win in the market

4.1 What Are Value Streams?

Before an understanding of value stream mapping can be achieved, there must be an understanding of value streams, what they represent, and their boundaries. "Value" is an activity that transforms the product in a way the customer is willing to pay for. A value stream is all the actions (both value added and nonvalue added) currently required to bring a product through the main flows essential to every product, which includes:

- Production: The production flow from raw material to customer
- New product development: The design flow from concept to launch

A value stream is an end-to-end collection of activities that create or achieve a result for a customer of the manufacturer. The value stream is made up of everything that supports the value stream tasks and activities, including the:

- People who perform the tasks and their knowledge and skills
- Tools and technology that are used to perform and support the value stream tasks
- Physical facilities and environment in which the value stream resides
- Organization and culture of the manufacturer that owns the value stream

- Values and beliefs that dictate the corporate culture and behaviors, and affect the way in which work is accomplished
- Communications channels and the way in which information is disseminated through the manufacturer
- Policies, procedures, and processes that govern the activities of the value stream
- Social systems that support the value stream

As described in the next section, value stream mapping focuses on the activities of the manufacturer's value streams. It takes a rapid, high-level look at the business within the context of the value streams. It focuses on rapidly identifying business problems and opportunities for improvement, and defining solutions to resolve those problems and to take advantage of identified opportunities to effect improvement. Through value stream mapping, an unbiased understanding of the business, the vision, and the strategies upon which the business is founded, the way it operates, and its strengths and weaknesses are developed. The results of value stream mapping focus on the need for business improvement through change, defining the changes necessary to resolve business problems, and the ways to make that change happen.

4.2 What Is Value Stream Mapping?

Value stream mapping (VSM) is a Lean manufacturing technique used to analyze the flow of materials and information currently required to bring a product or service to a consumer. The trend in today's competitive environment is to provide high-quality and low-cost products based on the voice of customers (VOC). To become a profitable company, an effective simplification of workflow and elimination of waste must be the targets for all manufacturers. VSM allows manufacturers to understand where they are and what wastes need to be eliminated. In another words, VSM helps us see value, waste, and also the sources of waste in a value stream. VSM is a valuable tool for manufacturers of all types.

VSM is the process of mapping the material and information flow of all components and subassemblies in a value stream that includes manufacturing, suppliers, and distribution to the customer. VSM is a hands-on tool to show workflow, information flow, and value using process cycle time and first time quality metrics of percent complete and percent accurate. VSM provides manufacturers a comprehensive understanding of waste associated with manufacturing operations. The objective is to create a picture of the system of processes from beginning to end. VSM helps to improve the value-added process through step-by-step review and identification of connections, activities, information, and flow. It provides a system perspective to increase value and eliminate waste.

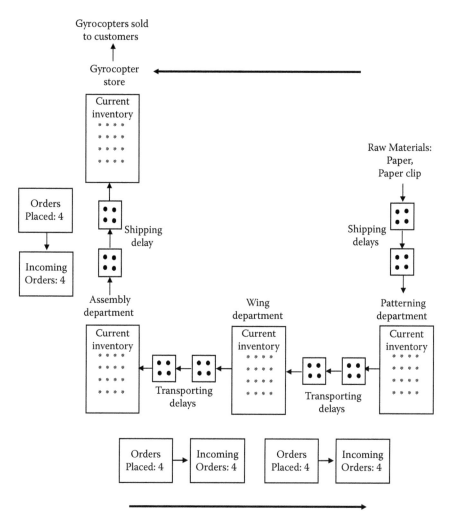

FIGURE 4.1
Value stream mapping for manufacturing gyrocopters.

There are two major implications in VSM that center on the involvement of key senior resources and the reuse of corporate assets. VSM focuses on the business, the vision, the direction, and how to achieve them. VSM projects are strategically oriented. They affect long-term business strategies and practices. This requires the involvement of key executive resources and decision makers within the manufacturer and comparable senior resources to manage and conduct the project activities.

VSM projects are completed within a matter of weeks. The rapid pace makes VSM dependent on the concepts of reuse. There are different types of reusable assets that expedite the delivery of VSM while minimizing risk. These include:

- Reuse of knowledge assets, such as technical templates, business templates, process templates, and package software
- Reuse of skill assets, which are those skills with an industry focus and those skills with a technology focus
- Reuse of process assets, including methodology, task structure and work plans, metrics, techniques, and iterative delivery

4.3 How to Complete Value Stream Mapping

VSM is an appropriate tool for identifying waste and, as such, many companies start by mapping their current processes. VSM is a map that outlines the current and future states of a production system, allowing users to understand where they are and what wastes need to be eliminated. In other words, it is used to see value, waste, and also the sources of waste in a value stream. VSM helps us to visualize more than just the single process level.

It is important to focus on one product family before starting the mapping because the customers care about their specific products and it is too complicated to include all product flows on one map. Once the value is specified from the customer perspective, value stream is identified for the product family. VSM involves walking and drawing the processing steps, both material and information, for one product family from door-to-door in the plant. VSM is primarily used to depict current and future, or "ideal," states in the process of developing implementation plans to implement Lean manufacturing. VSM is executed in the following five steps.

1. Clarify process to improve—We aim to improve the process of:

   Beginning of process:

   End of process:

   Checklist:

   a. Clarify aim
   b. Create high-level flowchart of process
   c. Add information and data flows
   d. Identify customer and supplies handoffs
   e. Perform observational walk
      - Note customer/supplier
      - Measure time of each step and total cycle time of process
   f. Determine delivery and quality requirements
   g. Design Lean/improved process

2. Make a high-level flowchart of the process. Include all steps required to deliver a service or product. Focus on the current process of how work is done. A second option is to create a deployment flowchart that clarifies roles and functions.

3. Identify and note customer and supplier connections for each step in diagram. Answer the following questions with your Lean manufacturing team.

   - What is the customer's need?
   - Who supplies what to whom?
   - How does each customer make a request?
   - How does each supplier respond?
   - How does a supplier do his or her other work?
   - What problems exist and what problems are solved?
   - By whom, when, where, and how?

4. Describe delivery and quality requirements. Determine key quality indicators.

5. Perform a value-stream waste observational walk through the process steps:

   - Follow movement of patient or product
   - Note information flow (paper, verbal, electronic)
   - Note inventory
   - Identify how work is "triggered" in the value stream
   - Identify how each step knows what to do next (sequencing)
   - Calculate process time, wait time, and first time quality for process steps and the entire value stream cycle (percent complete, percent accurate, number defect free)

The objective of value stream mapping is to rapidly identify business problems within the context of a value stream and to create solutions to resolve them. Value stream mapping is the process of working with customers to identify and assess the end-to-end collection of activities that comprise the manufacturer's value streams and create results for the customers of the value stream. The customer may be external to the manufacturer or the internal end user of the value stream product or result. The customer of the value stream is the recipient of the value produced by the value stream, regardless of active or passive involvement by the customer. Succinctly described, value stream mapping involves the following activities:

- Assessing the business vision and determining if it is sufficiently robust to carry the enterprise forward, creating a new or enhanced vision, if necessary

TABLE 4.1

Value Stream Mapping Planning Sheet

Current State Metrics	Current	Target
Process time (cycle time)		
Wait time		

First Time Quality
  % Accurate
  % Complete
  # of Defects in process

Create the Future State
  A. Use Lean principles and four rules for design to design an improved flow and process based on the waste you have identified.

  Lean Principles
  • Do work on time
  • Identify problems before it's too late
  • Eliminate waste
  • Reduce reproduction
  • Smooth workloads with standardized processes
  B. What are the customer requirements?
  C. Where and how will you trigger or sequence work?
  D. How will you make work flow smoothly? (Reduce interruptions due to handoffs, delays, queue or rework)
  E. How will work progress, or delays and problems be evident? What will you measure? Who will measure?

- Evaluating the alignment of executive management with the future vision
- Looking at the manufacturer to define the value streams that comprise it
- Determining the strengths and weaknesses of the value streams, based on analysis and feedback from external and internal sources
- Defining which value stream(s) will benefit from change and building a solution to effect the change within the context of the value stream(s)
- Creating a prioritized change management plan, identifying specific change projects that will achieve the desired level of improvement

See Table 4.1 for a value stream mapping plan.

4.4   Guidelines for Developing Value Stream Mapping

Value stream mapping begins at the level of the door-to-door flow in the plant. The steps of VSM are shown in Figure 4.2. Instead of recording each and every individual processing step, the process categories are drawn.

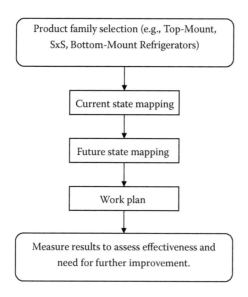

FIGURE 4.2
Value stream mapping steps.

Downstream process is the starting point of the mapping and work continues upstream. The current state is the mapping by collecting and using the actual data. The problems are identified and the solutions are decided by analyzing the current state map.

Furthermore, pacemaker and supermarket processes are determined, product mix is leveled, and the future state is mapped. The following data are needed to decide the future state:

- Cycle time
- Changeover time
- Production batch sizes
- Number of product variations
- Number of operators
- Pack size
- Working time (minus breaks)
- Scrap rate
- On-demand machine uptime

The arrows between current and future state go both ways in Figure 4.2 because the development of current and future states are overlapping efforts. Beyond these activities, a yearly value stream plan is created. This plan describes how to plan the transition from current state to future state. Then,

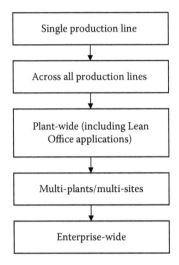

FIGURE 4.3
Levels of mapping the value streams.

as future state becomes reality, the mapping process repeats itself, because it always needs a future state. That is continuous improvement at the value stream level. The levels of VSM are shown in Figure 4.3.

Value stream viewpoint requires working with the big picture and improving the whole, rather than dealing with and improving each process separately. VSM is a way to plan Lean manufacturing by designing the whole door-to-door flow. It helps identify nonvalue-added steps, procurement time, distance traveled, and inventory level. What is meant by VSM is to draw, with symbols, each process in the material and information flow by following the production line from customer to supplier. Then, by asking a set of critical questions, future state of the flow showing how the flow should be is drawn. Here are four rules for developing value stream mapping:

1. All work must be highly specified as to content, sequence, timing, location, and expected outcome.
2. Every customer–supplier connection must be highly specified, direct, and there must be an unambiguous yes or no way to send requests and receive responses.
3. The pathway for every product and service must be predefined, highly specified, simple, and direct with no loops or forking.
4. Any improvement must be made in accordance with the scientific method by those closest to the work toward the ideal state.

In summary, the objective of VSM is to rapidly identify business problems within the context of a value stream and to create solutions to resolve them. VSM has two components: mapping the big picture and detailing the map.

4.5   Kaizen Event Analysis for Value Stream Mapping

Kaizen means "improvement." Kaizen event analysis is used to identify customer–enterprise interactions and responses. It is used in conjunction with customer value stream interaction analysis and customer satisfier analysis to fully define the profile of an enterprise's value streams.

Kaizen event analysis focuses on key events that are triggered by the customers of the value stream (e.g., request for information, placement of an order, etc.) or are triggered by the enterprise (e.g., marketing mailing, delivery or product, invoicing, etc.) for the benefit of the customer of the value stream. It is important to determine the initiator of the events and the typical response(s) to the event. This aids in identifying the activities that are considered value added and are involved in providing satisfaction.

In addition to defining the enterprise's value streams, event analysis is commonly used in developing workflows and process models. For these purposes, all events are identified and tracked to determine the sequence and timing of workflow steps and the dependencies between activities. In this context, an event may be classified as one of the following types:

- Request (customer orders goods)
- Passage of time (renew license on expiration date)
- Activity status change (issue purchase order finished)
- Object state change (stock drops below reorder point)

4.5.1   Purpose

The purpose of event analysis is to identify all key customer–enterprise interactions and responses to define the scope of the value stream(s) of the enterprise, and to understand the natural dependencies of business events to create detailed process models and workflows to enable a more effective information system design.

4.5.2   Benefits

The benefit of using event analysis is that it provides a structured approach to detailing the events that comprise a value stream, process, or workflow to create a more accurate and effective model or design.

4.5.3 Kaizen Event Analysis Guideline

Begin to explore customer–enterprise interactions by confirming the customer and value stream for which event analysis is to be applied. This may be most effectively accomplished in a workshop setting. Start by identifying all possible interactions between the customer and the value stream. To aid in focusing and structuring the event analysis, take a lifecycle view of the value stream. Highlight key events. Identify the direction and the typical response for all interactions. Discuss and explore each key event to ensure complete understanding of the triggers and the events that they initiate. If the responses to a single event vary widely, highlight the event as a potential area for focus. Try to resolve them before moving forward.

Identify any dependencies between the events, the interactions, and the responses. Include informational dependencies. For example, in the insurance industry, an event might be an automobile accident; the interaction between the customer and the enterprise, however, might be a phone call to the insurance company, submission of an accident report, or even a claim (depending on how the business is structured). The interaction between the customer and the enterprise and the response to the event may be dependent on information regarding the accident and the policy coverage terms. (For example, under the policy, the customer/claimant may be required to first notify the enterprise or insurance carrier before a claim can be filed. The policy may also dictate that the enterprise makes a site visit within four hours of the reported accident.)

Complex business environments may require additional exploration to complete all the details in event analysis. The focus during initial exploration and scoping of the value stream should be on documenting the key areas. The details can be explored more fully at later stages. Use work flow diagramming or dependency analysis to flesh out and complete the details of event analysis.

- Is the anticipated field of view manageable?
- What are the current problems with the value stream for this area of study (from the organization's perspective and from the customer's perspective)?
- What is the extent of variation in the area of study?
- What data is currently collected to measure activities in or about the area of study?
- Who touches the value stream?
- What is in and out of scope for the proposed VSM workshop?
- Who is directing the process?
- What do your customers want that you are currently not able to supply?
- Is there senior executive leadership support for this area of study?

- Is there sufficient funding available to support the VSM for this area of study?
- What is the anticipated schedule for the workshop(s)?

4.6 Business Bottom-Line Analysis of Value Streams

Despite the dramatic improvements that Lean thinking creates on the shop floor, many companies fail to see corresponding impacts on their financial statements. Implementing business bottom-line analysis that includes replacing the standard costing system with value stream costing helps to bridge this frustrating disconnect between shop-floor improvements and the financial statement. Such a business bottom-line analysis is effective to manage and control operations.

From the Lean manufacturing perspective, standard cost systems encourage people to do the wrong things. In basketball, an air ball is any shot that misses the basket completely and does not hit either the net, rim, or backboard. Some of a standard cost system's "air balls" include:

- Inducing companies to build in large batches to keep machines running and people busy to absorb overhead
- Not providing daily performance information needed to support continuous improvement at the cell and value stream levels
- Requiring huge amounts of computer resources and nonvalue-creating activities to gather and process work-order transactions
- Failing to capture the vast benefits that Lean generates in the form of higher quality, shorter lead times, and especially the added capacity gained from people, equipment, and space

The reason for the disconnect between Lean manufacturing and standard costing is that standard costing was developed as part of traditional accounting systems to support mass production. Such systems assume companies make money by keeping machines and people running in overdrive, creating mountains of widgets in order to lower unit costs. The standard costing system generates thousands of reports comparing the mountain's actual results against a standard in such areas as labor and machine utilization. Unfortunately, the reports are useless for managers trying to run and improve the manufacturing business.

As a Lean transformation of the shop brings processes under control and increases the velocity of material and information, the need for the transactions and complex and lengthy reports goes away. Batch production gives way to making today what the customer wants today. The mountain of widgets disappears. Inventory reduces because material is moving much faster.

The focus becomes optimizing the value stream for a product family, not optimizing individual operation. Supporting value-stream performance assessment and planning, business bottom-line analysis looks like this:

1. Developing cell performance measures
2. Establishing value stream performance measures
3. Setting up a value stream profit-and-loss (P&L) statement
4. Calculating improvements in capacity and turning this into valuable data for decision making
5. Building a value stream scorecard that integrates operational, capacity, and financial data

4.6.1 Value Stream Measures

Obviously, value stream mapping is the prerequisite to implement business bottom-line analysis including daily performance measures and weekly value stream measures. On the high-volume line, for example, the performance measures could include setup times and units produced every two hours as measured against the demand rate for the two-hour period. Performance for both measures is recorded on large dashboards near the two cells in the value stream. Every two hours, a supervisor records the cell's output and initials the board. If production is missed, the supervisor lists the reason and the countermeasures taken, and then initials the board. Value stream performance is recorded monthly (weekly is the goal) in areas such as:

- Customer service, measured as on-time delivery to the customer's requested delivery date
- Productivity, measured as sales dollars per person
- Inventory, measured as the number of "days of inventory" in the value stream
- Safety, measured as lost-time accidents
- Quality, measured as scrap

Results for each measure are graphed and posted on the wall near the value stream's end. Below these hang charts showing the top three problems in the areas of customer service, productivity, inventory, safety, and quality. Below these are lists of ideas for solving the problems.

- In requirement 1, the costs for each chair include only manufacturing costs.
- In requirement 2, the costs for each chair include manufacturing costs as well as upstream costs and downstream costs.

It is important for Schramka Company to take into account other than manufacturing costs for strategic decisions, especially long-term pricing

Activity-based job costing.

			Executive Gyrocopter	Private Gyrocopter
1. Direct manufacturing costs				
	Direct materials		$600,000	$25,000
	Direct manufacturing labor			
		$20 × 7,500; $20 × 500	150,000	10,000
		Direct manufacturing costs	750,000	35,000
Indirect manufacturing costs				
	Materials handling			
	Cutting	$0.25 × 100,000; $0.25 × 3,500	25,000	875
	Assembly	$2.50 × 100,000; $2.50 × 3,500	250,000	8,750
		$25.00 × 7,500; $25.00 × 500	187,500	12,500
		Total indirect manufacturing costs	462,500	22,125
Total manufacturing costs	$1,212,500	$57,125		
Unit Costs				
		Executive gyrocopter: $1,212,500 ÷ 5,000 = $242.50		
		Private gyrocopter $57,125 ÷ 100 = $571.25		
			Executive Gyrocopter	Private Gyrocopter
2				
	Upstream costs		$60.00	$146.00
	Manufacturing costs		242.5	571.25
	Downstream costs		110	236
	Total costs		$412.50	$953.25

FIGURE 4.4
Activity-based job costing.

decisions and product emphasis. When comparing the Executive Gyrocopter and the Private Gyrocopter, the Private Gyrocopter uses more of the upstream and downstream activities per unit than does the Executive Gyrocopter. The Private Gyrocopter also uses more of the manufacturing activities per unit than the Executive Gyrocopter.

4.6.2 Traditional Product-Costing Systems: Problems with Overhead

One of the factors in establishing prices (what your customer pays you) is the cost of producing your product or service. If you think that your cost is higher than it really is, you might set your price too high and lose customers. If you think that your cost is lower than it really is, you might set your price too low and not make enough profit (it is possible that your price could be so low that you are not even covering your cost). Obviously, knowing the real cost of the product or service you are producing is important. An overview of the traditional product-costing system is shown in Figure 4.5. Problems with overheads embedded in traditional product-costing system include:

- The best assignment of costs occurs when the costs are traceable directly to the product produced or the service rendered. Direct costs (direct material and direct labor) are not a problem because these costs can be traced to a single cost objective (product or service) at reasonable cost.

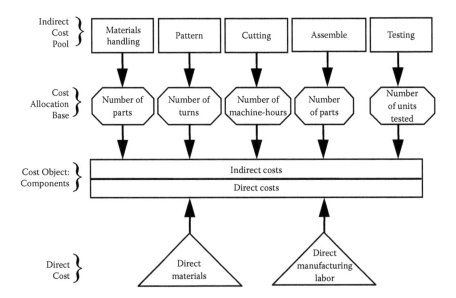

FIGURE 4.5
Overview of the traditional product-costing system.

Traditional Product-Costing Method	Job Order 110	Job Order 111
	Patterning	Processing
Direct manufacturing costs:		
Direct materials	$9,700	$59,900
Direct manufacturing labor	750	11,250
$30* 25; $30* 375	$10,450	$71,150
Indirect manufacturing costs		
$115* 25; $115* 375	2,875	43,125
Total manufacturing costs	$13,325	$114,275
Number of units	÷ 10	÷ 200
Manufacturing costs per unit	$1,332.50	$571.38

FIGURE 4.6
Traditional product-costing calculation for patterning and cutting of a gyrocopter.

- Indirect costs (manufacturing overhead) are common costs that cannot be directly traced to individual products or services. Often the most difficult part of computing accurate unit costs is determining the proper amount of overhead cost to assign to each product, service, or job.
- Traditional costing systems assigned all manufacturing overhead using the same predetermined overhead rate. This often caused too much overhead to be assigned to some products and too little overhead to be assigned to other products.

Product costs are generally defined as follows:

Product cost = Direct material + Direct labor + Factory overhead

Or on a per unit basis:

Product cost per unit = Direct material per unit + Direct labor per unit
+ Factory overhead per unit

Figure 4.6 shows a traditional product-costing calculation for patterning and cutting of a gyrocopter.

4.6.3 Activity-Based Costing (ABC)

Activity-based costing (ABC) is a system that first accumulates manufacturing overhead (indirect) costs for each of the activities of an organization, and then assigns the costs of activities to the products, services, or cost objects that caused the activity. ABC allocates (or assigns) overhead to multiple cost pools (a cost pool for each activity). A traditional costing system uses one cost

Activity-Based Costing Method	Job Order 410 Patterning	Job Order 411 Processing
Direct manufacturing costs:		
Direct materials	$9,700	$59,900
Direct manufacturing labor	750	11,250
$30 × 25; $30 × 375	$10,450	$71,150
Indirect manufacturing costs:		
Materials handling	200	800
$0.40 × 500; $0.40 × 2,000		
Lathe work	4,000	12,000
$0.20 × 20,000; $0.20 × 60,000		
Milling	3,000	21,000
$20.00 × 150; $20.00 × 1,050		
Grinding	400	1,600
$0.80 × 500; $0.80 × 2,000		
Testing	150	3,000
$15.00 × 10; $15.00 × 200	7,750	38,400
Total manufacturing costs	$18,200	$109,550
Number of units per job	÷ 10	÷ 200
Unit manufacturing cost per job	$1,820	$547.75

FIGURE 4.7
Activity-based costing calculation for patterning and cutting of a gyrocopter.

pool for all manufacturing overhead costs. ABC assigns the costs in the cost pools (overhead costs) to products using different cost allocation bases (called cost drivers) for the different cost pools. A traditional costing system uses one cost pool with one cost allocation base. The reasoning behind ABC is simple: products consume activities; activities consume resources, which have costs.

ABC was developed to get more accurate costs for products, services, or cost objects by assigning the right amount of manufacturing overhead to the product, service, or cost object. Figure 4.7 shows an activity-based costing calculation for patterning and cutting of a gyrocopter.

ABC is a costing method that is designed to provide Lean manufacturing practitioners with cost information for strategic and other decisions that potentially affect capacity and therefore "fixed" costs. Reasons for ABC include:

- Fierce competitive pressure has caused the need to know profit margins for individual products and services.
- Businesses have more products, services, and customer classes with greatly differing consumption of shared resources.
- New production techniques have increased the proportion of indirect costs that must be allocated correctly.

- Technological change has shortened product life cycles reducing the time to correct wrong costs or prices.
- Big problems (bids lost due to overpriced products; prices that are too low on underpriced products) result from bad decisions based on inaccurate cost determination.
- Computer technology has reduced the costs of developing and operating activity-based costing systems.

4.6.4   How Is Activity-Based Costing Different from Traditional Product Costing

In the traditional cost accounting system, the objective is to properly value inventories and cost of goods sold for external financial reports. In ABC, the objective is to understand overhead and the profitability of products and customers. The major differences are:

1. Nonmanufacturing as well as manufacturing costs may be assigned to products.
2. Some manufacturing costs may be excluded from product costs.
3. There are a number of overhead cost pools, each of which is allocated to products and other costing objects using its own unique measure of activity.
4. The allocation bases often differ from those used in traditional costing systems.
5. The overhead rates, or activity rates, may be based on the level of activity at capacity rather than on the budgeted level of activity.

In traditional cost accounting, only manufacturing costs are assigned to products. Selling, general, and administrative expenses are treated as period expenses and are not assigned to products. Under ABC, however, nonmanufacturing costs such as commissions paid to salespersons, shipping costs, and warranty repair costs, which can be easily traced to individual products, are included in the product costs. On the other hand and despite the traditional cost accounting, some manufacturing costs that are not caused by the products (i.e., security guard's wage) do not participate in product costing under ABC. A cost is assigned to a product only if there is good reason to believe that the cost would be affected by decisions concerning the product. Figure 4.8 compares traditional product costing and ABC for patterning and processing of a gyrocopter.

Job order 110 has an increase in reported unit cost of 36.6 percent [($1,820 − $1,332.50) ÷ $1,332.50], while job order 411 has a decrease in reported unit cost of 4.1 percent [($547.75 − $571.375) ÷ $571.375].

A common finding when ABC is implemented is that low-volume products have increases in their reported costs while high-volume products have

Traditional Product-Costing vs. Activity-Based Costing	Job Order 110 Patterning	Job Order 111 Processing
Number of units in job	10	200
Costs per unit with prior costing system	$1,332.50	$571.38
Costs per unit with activity-based costing	1,820.00	547.75

FIGURE 4.8
Comparison of traditional product costing and activity-based costing for patterning and processing of a gyrocopter.

decreases in their reported cost. This result is also found in requirements 1 and 2 of this problem. Costs such as materials-handling costs vary with the number of parts handled (a function of batches and complexity of products) rather than with direct manufacturing labor hours, an output-unit level cost driver, which was the only cost driver in the previous job-costing system.

The product cost figures computed in requirements 1 and 2 differ because the job orders differ in the way they use each of the five activity areas, and the activity areas differ in their indirect cost allocation bases (specifically, each area does not use the direct manufacturing labor-hours indirect cost allocation base).

In traditional cost accounting, predetermined overhead rates are computed by dividing budgeted overhead costs by a measure of budgeted activity such as budgeted direct labor hours. This practice results in applying the costs of unused, or idle, capacity to products, and it results in unstable unit product costs. In ABC, however, products are charged for the costs of capacity they use. In other words, the costs of idle capacity are not charged to products. This results in more stable unit costs and is consistent with the objective of assigning only those costs to products that are actually caused by the products. Figure 4.9 documents how the two job orders differ in the way they use each of the five activity areas included in indirect manufacturing costs.

	Usage Based on Analysis of Activity Area Cost Drivers		Usage Assumed with Direct Manuf. Labor-Hours as Application Base	
	Job Order 110 Patterning	Job Order 111 Processing	Job Order 110 Patterning	Job Order 111 Processing
Activity Area				
Materials handling	20.00%	80.00%	6.25%	93.75%
Lathe work	25	75	6.25	93.75
Milling	12.5	87.5	6.25	93.75
Grinding	20	80	6.25	93.75
Testing	4.8	95.2	6.25	93.75

FIGURE 4.9
How the two job orders differ in the way they use each of the five activity areas included in indirect manufacturing costs.

The differences in product cost figures might be important to Marion Avionics Corporation (MAC) for product pricing and product emphasis decisions. The ABC approach indicates that job order 110 is being underpriced while job order 111 is being overpriced. MAC may erroneously push job order 110 and deemphasize job order 111. Moreover, by its actions, MAC may encourage a competitor to enter the market for job order 111 and take market share away from it.

Information from the ABC system can also help MAC manage its business better in several ways.

1. Product design. Product designers at MAC likely will find the numbers in the ABC approach more believable and credible than those in the existing system. In a machine-paced manufacturing environment, it is unlikely that direct labor hours would be the major cost driver. ABC provides more credible signals to product designers about the ways the costs of a product can be reduced—for example, use fewer parts, require fewer turns on the lathe, and reduce the number of machine hours in the milling area.

2. Cost management. MAC can reduce the cost of jobs by making process improvements that reduce the activities that need to be done to complete jobs and by reducing the costs of doing the activities.

3. Cost planning. ABC provides a more refined model to forecast costs and to explain why actual costs differ from budgeted costs.

Activity-based management (ABM) is an extension of ABC from a product costing system to a management function that focuses on reducing costs and improving processes and decision making. ABM classifies activities as

- Value-added activities—Activities that increase the worth of a product or service (e.g., engineering design, machining, assembly, painting, and packaging)
- Nonvalue-added activities—Activities that add costs to or increase the time spent on a product or service without increasing its market value (e.g., inspecting, moving, and storing inventories of materials; work in process and finished goods)

Certain nonvalue-added activities are not totally wasteful and cannot be totally eliminated, but these activities should be minimized. The primary benefit of ABC is more accurate product costing because

- ABC leads to more cost pools.
- ABC leads to enhanced control over overhead costs.
- ABC leads to better Lean manufacturing decisions.

ABM focuses on managing activities as a way of eliminating waste and reducing delays and defects. In this regard, powerful tools such as total quality management, process reengineering, and benchmarking can be used, which provide a systematic approach to identifying the activities with greatest room for improvement.

Implementing ABC often will typically shift costs from high-volume to low-volume products, but the effects will be much more dramatic on the unit costs of the low-volume products. The unit costs of the low-volume products will increase far more than the unit costs of the high-volume products will decrease.

5

From Lognormal to Cobb–Douglas
Distribution: Lean Production Analysis

This chapter enables you to

- Explain the concepts of a firm's technology and associated production function
- Distinguish between the short-run and long-run production periods
- Derive a short-run production function
- Explain and calculate the marginal physical product of an input
- Distinguish between increasing, constant, and diminishing returns to a factor

---

5.1 Technology: The Production Function

A production function asserts that the maximum output of a technologically determined production process is a mathematical function of input factors of production. A firm's option for combining inputs into outputs is called its *technology*. The firm's technology is summarized by its production function. A *production function* shows the maximum output that can be produced from any given combination of inputs:

$$Q = F(L, K)$$

where
  Q = Units of output
  L = Units of labor input
  K = Units of capital input

Thus, a production function can be described as the specification of the minimum input requirements needed to produce designated quantities of output, given available technology.

    Why maximum output? Because it is assumed that a manufacturer uses only technologically efficient methods. By assuming that the maximum output technologically possible from a given set of inputs is achieved,

TABLE 5.1

Example for a Production
Function

Labor			
5	30	34	37
4	26	30	33
3	21	25	28
2	16	20	23
1	10	13	15
	1	2	3
			Capital

production analyses are abstracting away from the engineering and managerial problems inherently associated with a particular production process. This implies that using more of one input and either the same amount or more of the other input must increase output.

A production function can be represented in Table 5.1. In this table five units of labor and two of capital can produce thirty-four units of output. It is, of course, always possible to waste resources and to produce fewer than thirty-four units with five units of labor and two of capital, but the table indicates that no more than thirty-four can be produced with the technology available. The production function thus contains the limitations that technology places on the firm.

5.2   Long-Run and Short-Run Production Functions

The production possibilities available to a firm depend on the time horizon over which the firm makes its input choices. A decision frame, in which one or more inputs are held constant, may be used; for example, capital may be assumed to be fixed or constant in the short run, and only labor variable, while in the long run, both capital and labor factors are variable. In general, the more time a firm has to make its decisions, the more input choices it has. In particular, an important distinction exists between the long-run and short-run production periods.

- The *short run* is defined as the period of time during which the quantity of only *one* of the firm's inputs can be varied; all the others are fixed.
- The *long run* is defined as the period of time long enough for the quantity of *all* the firm's inputs to be varied.

An input (factor) whose level can be varied over the relevant time period is known as a variable factor. An input that cannot be varied is called a fixed factor.

5.2.1   Long-Run Production Function

$Q = f(K, L)$. Quantity is a function of the inputs used to produce it; in this example capital and labor. Quantity is measured as a rate of production (flow) as are capital and labor, for example, the amount of cars washed per day is a function of the amount of labor and capital used each day.

- The production function specifies a technically efficient use of labor and capital necessary to produce output, that is, no resources are "wasted."
- The cost function specifies an economically efficient use of resources, that is, the firm chooses the least cost combination of inputs, to produce a given output.
- This yields the long-run cost function: total costs $(C) = g(Q)$, which depends on the prices of inputs. This function can be a pretty good proxy for the opportunity cost of delivering Q, at least where we measure costs in units of present value: that is, the change in present value or owner's equity caused by some specified action (and where for purposes of measurement the attendant increase in wealth is excluded from the computation of equity).

Long-run functional relationships are traditionally obtained in the single product case: cost varies as a function of total production volume (V), the rate of output (x), the date of first delivery (T), and the date of completion of the full production run (m), where $x(t)$ denotes the rate of output at moment t. Moreover, as the total quantity of units produced increases, the cost of future output tends to decline because production knowledge increases as a result of production experience (this proposition is known as the learning or progress curve). That is

$$dC/dT \mid x = x_0, \quad V = V_0, < 0 \tag{5.1}$$

This relationship probably holds for most products produced in large batches using traditional mass-production methods. Moreover it is usually assumed that

$$dC/dx(t) \mid T = T_0, \quad V = V_0 > 0 \tag{5.2}$$

$$d^2C/dx(t)^2 \mid T = T_0, \quad V = V_0 > 0 \tag{5.3}$$

$$dC/dV \mid x = x_0, \quad T = T_0 > 0 \tag{5.4}$$

$$d^2C/dV^2 \mid x = x_0, \quad T = T_0 < 0 \tag{5.5}$$

Many of these functional relationships have been attenuated by the rise of computer-assisted design and manufacturing technology and modern information.

5.2.2 Short-Run Production Function

In a short-run production function at least one of the X's (inputs) is fixed. In the long run, all factor inputs are variable at the discretion of management. The short-run production function shows the maximum output the firm can produce when only one of its inputs can be varied, the others remaining fixed. Assuming two inputs, capital and labor, and that capital is fixed and labor is the variable factor, we can write the short-run production function as follows:

$$Q = F(L, \bar{K}) \tag{5.6}$$

where the bar over the symbol for capital units indicates that it is a constant amount.

This relationship between the variable input (e.g., labor) and total output constitutes the firm's short-run total product curve. In particular we are interested in how much total output increases when the firm uses more of the variable input.

EXAMPLE 5.1

A short-run production function for paper gyrocopters can be represented in Table 5.2. As the firm increases the amount of labor it employs with its fixed quantity (30 units) of capital, the maximum total output produced increases. But notice that it does not increase at a steady rate throughout.

As shown by Figure 5.1, total paper gyrocopters initially increases quite quickly (the total product curve gets steeper). Then it increases at steady rate (the slope of the total product [TP] curve is constant). Finally, TP rises at a slower rate (the slope gets flatter). This pattern in the rate of increase in total output is measured by the *marginal physical product* of labor. In the short run, some inputs cannot be varied, so the firm does not have as much flexibility as in the

TABLE 5.2

Short-Run Production Function

Units of Capital Per Day	Units of Labor Per Day	Total Paper Gyrocopters Produced Per Day	Marginal (Physical) Product
30	0	0	
30	1	10	10
30	2	22	12
30	3	37	15
30	4	52	15
30	5	67	15
30	6	78	11
30	7	86	8

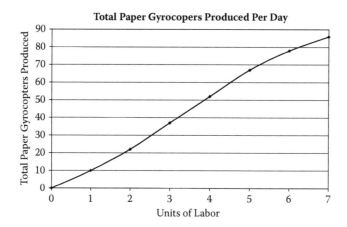

FIGURE 5.1
Short-run production function for paper gyrocopters.

long run. In this case, the short-run production function is a function of only the inputs that can be varied. Suppose that capital is fixed in the short run. Then $Y = g(labor)$.

5.3   Three Stages of Production

Let L equal number of units of labor input and TP the number of units of product/output, then average product (AP) can be calculated by

$$AP = \frac{TP}{L} \tag{5.7}$$

AP, the output per unit of labor input, is frequently reported in press.
   Marginal product (MP) can be calculated by

$$MP = \frac{\Delta TP}{\Delta L} \tag{5.8}$$

MP, the output attributable to last unit of labor used, is what manufacturing engineers think of. The MP of an input is the extra amount of output that can be produced when the firm uses one additional unit of an input, holding all other inputs constant.
   The usual shapes of the short-run production function are

1. Increasing—Demonstrates increasing marginal returns, that is, as additional units of input are added to the production process, each additional unit of input adds more and more to the number of units of output/product produced. A short-run production function with

TABLE 5.3

Increasing Marginal Returns

L	TP	MP	AP
0	0		
1	10	10	10
2	30	20	15
3	60	30	20
4	100	40	25

increasing marginal returns for paper gyrocopters is represented in Table 5.3. Here *TP* is increasing at an increasing rate. That increasing rate means that the *MP* is increasing. In this stage of production all workers should be hired as each additional worker adds more than the previous worker to total production. In the short run, output at first increases at an increasing rate with increases in labor (increasing returns to labor as shown by Figure 5.2).

2. Linear—Demonstrates constant marginal returns, that is, as additional units of input are added to the production process, each additional unit of input adds an equal amount to the number of units of output/product produced. A short-run production function with constant marginal returns for paper gyrocopters is represented in Table 5.4. Then output increases at a constant rate with increases in labor (constant returns to labor as shown in Figure 5.3).

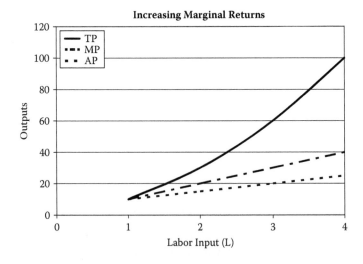

FIGURE 5.2
Short-run production function with increasing marginal returns.

TABLE 5.4

Constant Marginal Returns

L	TP	MP	AP
0	0		
1	10	10	10
2	20	10	10
3	30	10	10
4	40	10	10
5	50	10	10

3. Decreasing—Demonstrates diminishing marginal returns, that is, as additional units of input are added to the production process, each additional unit of input adds less and less to the number of units of output/product produced. A short-run production function with decreasing marginal returns for paper gyrocopters is represented in Table 5.5. Here *TP* is increasing at a decreasing rate. That decreasing rate means that the *MP* is decreasing. In this state of production, also known as "the economic region of production," the firm will locate the profit-maximizing amount of workers to hire. Finally, output increases at a decreasing rate with increases in labor (diminishing returns to labor as shown in Figure 5.4).

A typical production function includes all the three theoretical cases presented on the previous page. The firm, whose objective is to maximize profits,

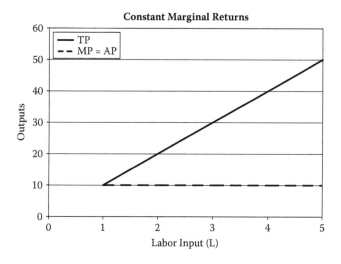

FIGURE 5.3
Short-run production function with constant marginal returns.

TABLE 5.5

Decreasing Marginal Returns

L	TP	MP	AP
0	0		
1	200	200	200
2	350	150	175
3	450	100	150
4	525	50	131.25

will operate in the area of diminishing returns. This result will be explained with additional comments from class lecture. As shown by Figure 5.5, short-run cost functions are larger than long-run cost functions because, in the short run, fixed inputs cannot be varied. In the long run, all inputs can be varied, and this greater flexibility allows you to achieve lower costs.

In manufacturing, a company would keep producing as long as the benefit of producing another unit exceeds the cost of producing another unit. As shown by Figure 5.6, you can see that average cost intersects the marginal cost curve at about eleven laborers (output ≈ 20).

As we've seen from Figure 5.6, the key concept with regard to the firm's short-run production function is *MP*. To use *MP* to describe the production function we need to know what happens to the *MP* of a factor as the firm increases its use of that factor. There are three possibilities:

- Increasing marginal returns—The *MP* of an input increases as more of the input is used.

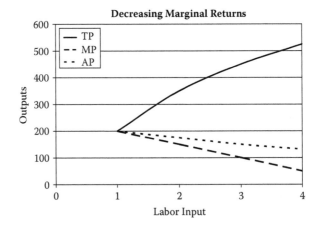

FIGURE 5.4
Short-run production function with decreasing marginal returns.

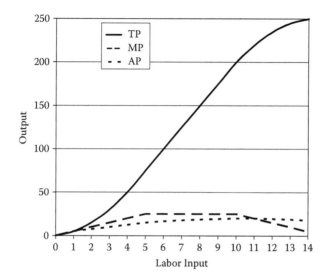

FIGURE 5.5
Usual shapes of the short-run production function.

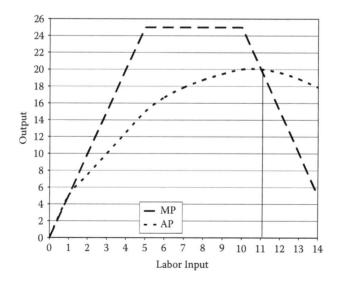

FIGURE 5.6
Determine optimal production.

- Constant marginal returns—The *MPP* of an input remains constant as more of the input is used.
- Diminishing marginal returns—The *MPP* of an input decreases as more of the input is used (i.e., the famous "law" of diminishing returns).

Our example short-run production function shows a typical case where there is increasing marginal returns to labor at first, then a period of constant returns, and eventually diminishing marginal returns to labor begin to occur.

5.4   Lognormal Distribution

A variable might be modeled as lognormal if it can be thought of as the multiplicative product of many small independent factors. Production functions help companies estimate how many goods can be produced with a given number of inputs. A commonly used production function is one of the form

$$Q = f(L, K) = aL^b K^c \tag{5.9}$$

which can be modeled as a lognormal distribution.

A random variable X is said to be lognormal distributed with parameters μ and σ^2 if

$$Y = \log X \tag{5.10}$$

is normally distributed with mean μ and variance σ^2. It is clear that X has to be a positive random variable. This distribution is quite useful in studying the behavior of stock prices.

For the lognormal distribution as described earlier, the probability density function (pdf) is

$$g(x) = \frac{1}{\sqrt{2\pi}\sigma x} e^{-\frac{1}{2\sigma^2}(\log x - \mu)^2}, \quad x > 0 \tag{5.11}$$

The lognormal distribution is sometimes called the antilognormal distribution, because it is the distribution of the random variable X that is the antilog of the normal random variable Y. However, "lognormal" is most commonly used in the literature. When applied to production data, it is often called the Cobb–Douglas distribution.

We next discuss some properties of the lognormal distribution, as defined in Equation 1. The rth moment of X is

$$\mu'_r = E(X^r) = E(e^{rY}) = e^{\mu r + \frac{r^2 \sigma^2}{2}} \tag{5.12}$$

It is noted that we have utilized the moment-generating function (MGF) of the normal random variable Y with mean μ and variance σ^2 is

$$M_Y(t) = e^{\mu t + \frac{t^2 \sigma^2}{2}} \qquad (5.13)$$

Thus, $E(e^{rY})$ is simply $M_Y(r)$, which is the right-hand side of Equation 3. From Equation 3 we have

$$E(X) = e^{\mu + \frac{\sigma^2}{2}} \qquad (5.14)$$

$$Var(X) = e^{2\mu} e^{\sigma^2} [e^{\sigma^2} - 1] \qquad (5.15)$$

Thus, the fractional and negative moment of a lognormal distribution can be obtained from Equation 5.13.

From Equation 5.13 we can also obtain the skewness and kurtosis of the lognormal distribution. We will first recall the following formulae,

$$\mu_3 = E(X - \mu_1')^3 = \mu_3' - 3\mu_1'\mu_2' + 2\mu_1'^3 \qquad (5.16)$$

and

$$\mu_4 = E(X - \mu_1')^4 = \mu_4' - 4\mu_1'\mu_3' + 6\mu_1'\mu_2'^2 - 3\mu_1'^4 \qquad (5.17)$$

Substituting μ_i' for $i = 1, 2, 3$, and 4, into Equations 5.15 and 5.17, we have

$$\begin{aligned}
\mu_3 &= (\mu_1')^3 [e^{3\sigma^2} - 3e^{\sigma^2} + 2] \\
&= (\mu_1')^3 [(e^{3\sigma^2} - 3e^{2\sigma^2} + 3e^{\sigma^2} - 1) + 3(e^{2\sigma^2} - 2e^{\sigma^2} + 1)] \\
&= (\mu_1')^3 (\eta^3 + 3\eta^2)
\end{aligned} \qquad (5.18)$$

and

$$\begin{aligned}
\mu_4 &= (\mu_1')^4 [e^{6\sigma^2} - 4e^{3\sigma^2} + 6e^{\sigma^2} - 3] \\
&= (\mu_1')^4 [\eta^6 + 6\eta^5 + 15\eta^4 + 16\eta^3 + 3\eta^2]
\end{aligned} \qquad (5.19)$$

where $\eta = e^{\sigma^2} - 1$.

It is noted that the moment sequence $\{\mu_r'\}$ does not belong only to lognormal distribution. Thus, the lognormal distribution cannot be defined by its moments. The cdf of X is

$$P(X \leq x) = P(\log X \leq \log x) = N\left(\frac{\log x - \mu}{\sigma}\right) \qquad (5.20)$$

because $\log X$ is normally distributed with mean μ and variance σ^2. The distribution of X is unimodal with the mode at

$$mode(X) = e^{(\mu - \sigma^2)} \qquad (5.21)$$

Let x_α be the $(100)\alpha$ percentile for the lognormal distribution and z_α be the corresponding percentile for the standard normal, then

$$P(X \le x_\alpha) = P\left(\frac{\log X - \mu}{\sigma} \le \frac{\log x_\alpha - \mu}{\sigma}\right) = N\left(\frac{\log x_\alpha - \mu}{\sigma}\right) \qquad (5.22)$$

Thus $z_\alpha = \frac{\log x_\alpha - \mu}{\sigma}$, implying that

$$x_\alpha = e^{\mu + \sigma z_\alpha} \qquad (5.23)$$

Thus, the percentile of the lognormal distribution can be obtained from the percentile of the standard normal.

From Equation 5.17, we also see that

$$median(X) = e^\mu \qquad (5.24)$$

as $z_{0.5} = 0$. Thus $median(X) > mod\ e(X)$. Hence the lognormal distribution is not symmetric.

5.5   Cobb–Douglas Production Function

As described in Section 5.3, production functions help companies estimate how many goods can be produced with a given number of inputs. A commonly used production function is one of the form $Q = f(L, K) = aL^b K^c$ and is called the Cobb–Douglas production function, named after the two founders. In the Cobb–Douglas production function, L is the number of units of labor, K is the number of unit of capital, and a, b, c are constants.

EXAMPLE 5.2

Suppose a metal manufacturing company has a Cobb–Douglas production function, in hundreds of pounds, of $f(L, K) = 10L^{0.25} K^{0.75}$ where L is the number of hours of labor and K is the dollar amount of capital invested. If the company uses 2000 hours of labor and $1,500 in capital, how many pounds of metal will be produced?

To find the number of pounds of metal produced we need to evaluate

$$f(2000, 750) = 10(2000)^{0.25}(1500)^{0.75}$$

$$= 10(6.6874)(241.0285)$$

$$\approx 16119$$

Thus, the manufacturing company will produce approximately 1,611,900 pounds of metal when 2000 hours of labor and $1,500 in capital is invested.

The advantages of using the Cobb–Douglas production function for estimating production are many and are widely used in empirical work. One advantage of the Cobb–Douglas function is that returns to scale can be determined by simply summing the exponents on the two inputs, labor and capital. Figure 5.7 shows how to determine the returns to scale.

Let's look at this in the form of an example. If a company's Cobb–Douglas production function is represented by $f(L, K) = L^{0.3} K^{0.7}$, then $f(2, 4) = 2^{0.3} 4^{0.7} = 3.249$. If we increase labor and capital by a factor of 3 then we would expect production to increase by a factor of 3 because $0.3 + 0.7 = 1$.

$$f(6, 12) = 6^{0.3} 12^{0.7} = 9.747$$

We notice that $3(3.249) = 9.747$. Thus, we have confirmed the production function displays constant returns to scale.

Now suppose a company has a Cobb–Douglas production function of $f(L, K) = L^{0.4} K^{0.8}$, then $f(2, 4) = 2^{0.4} 4^{0.8} = 4$. Once again, if we increase labor and capital by a factor of 3 then we would expect production to increase by more than a factor of 3 because $0.4 + 0.8 = 1.2$.

$$f(6, 12) = 6^{0.4} 12^{0.8} = 14.949$$

Since $14.949 > (4)(3) = 12$, the production function displays increasing returns to scale.

Last, suppose a company has a Cobb–Douglas production function of $f(L, K) = L^{0.2} K^{0.5}$, then $f(2, 4) = 2^{0.2} 4^{0.5} = 2.297$. If labor and capital are increased

---

Properties of the Cobb–Douglas Production Function

Given a Cobb–Douglas production function $f(L, K) = aL^b K^c$ if

- $b + c = 1$ then the function displays constant return to scale. That is, increasing labor and capital by a factor of p causes production to increase by the same factor p.
- $b + c > 1$ then the function displays increasing returns to scale. That is an increase in labor and capital by a factor p causes production to increase by more than the factor p.
- $b + c < 1$ then the function displays decreasing returns to scale. That is an increase in labor and capital by a given factor p causes production to increase by less than the factor p.

FIGURE 5.7
Properties of the Cobb–Douglas production function.

by a factor of 3, then we would expect production to increase by less than a factor of 3 because $0.2 + 0.5 = 0.7 < 1$.

$$f(6,12) = 6^{0.2}12^{0.5} = 4.957$$

Since $4.957 < (2.297)(3) = 6.891$ the production function displays decreasing returns to scale.

EXAMPLE 5.3

A sporting goods manufacturing company has a Cobb–Douglas production function of $f(L, K) = 2L^{.36} K^{.43}$, where $f(L, K)$ is the number of units produced. If the company invests 800 hours of labor and $3,000 in capital, how many sporting good units will be produced? What type of returns to scale does the production function display?
   Evaluating $f(800, 3000)$ we get

$$f(800, 3000) = 2(800)^{0.36} (3000)^{0.43} \approx 694$$

Thus, the sporting goods manufacturing company can expect to produce 694 units of sporting goods when 800 hours of labor and $3,000 in capital are invested. Since the sum of the exponents on the inputs are less than 1, that is $0.36 + 0.43 = 0.79$, the production function displays a decreasing returns to scale.

Returns to scale are important for determining how many firms will populate an industry. When increasing returns to scale exist, one large firm will produce more cheaply than two small firms. Small firms will thus have a tendency to merge to increase profits, and those that do not merge will eventually fail. On the other hand, if an industry has decreasing returns to scale, a merger of two small firms to create a large firm will cut output, raise average costs, and lower profits. In such industries, many small firms should exist rather than a few large firms.

Opportunities to increase profits from improving material, labor, and production processes are potentially much larger than for energy or waste reductions since material, labor, and production expenses typically comprise a larger part of the overall budget than energy or waste expenses. Savings opportunities can be identified by applying principles of Lean production such as minimal inventory, quick changeovers, one-piece flow, and preventive maintenance to a manufacturing process.

5.6   Quantifying Additional Profit from Productivity Gains

In many cases productivity recommendations result in increasing the quantity of products produced per unit time. One way to quantify the profit from this type of productivity gain is to break out the production expenses and calculate the remaining profit. To use this methodology, consider profit as the difference between sales revenue and the following production expenses: labor, materials, energy, overhead, and taxes. An example of the percentage of sales revenue associated with each item, and the actual amount is shown next for a company with sales revenue of $100,000 per year.

Sales revenue	100%	$100,000
Labor	30%	–$30,000
Materials	20%	–$20,000
Energy	2%	–$2,000
Overhead	8%	–$8,000
Taxes	30%	–$30,000
Profit	10%	$10,000

To quantify the additional profit from a productivity gain, it must be known whether the company could sell more products if it produced more products. Generic examples of quantifying additional profit from productivity gains in each case are given next.

5.6.1 Could Sell More Products if Produced More

Consider a recommendation resulting in a productivity gain of 10 percent for the aforementioned company. If the company could sell more products if it produced more products, the sales revenue would increase to $110,000 per year. To sell more products, materials, energy, and tax expenses would scale at the same percentage as before, while labor and overhead costs would remain constant. In this case, the profit would be

Sales revenue	100%	$110,000
Labor	27%	–$30,000
Materials	20%	–$22,000
Energy	2%	–$2,200
Overhead	7.3%	–$8,000
Taxes	30%	–$33,000
Profit	10%	$14,800

Thus, the increased profit from this productivity increase would be: $14,800 – $10,000 = $4,800.

5.6.2 Could Not Sell More Products if Produced More

Consider a recommendation resulting in a productivity gain of 10 percent for the aforementioned company. If the company could not sell more products if it produced more products, the sales revenue would remain $100,000 per year. Material, energy, tax, and overhead expenses would remain the same, while labor expenses would reduce. The profit would be

Sales revenue	100%	$100,000
Labor	27%	–$27,000
Materials	20%	–$20,000
Energy	2%	–$2,000
Overhead	8%	–$8,000
Taxes:	30%	–$30,000
Profit	13%	$13,000

Thus, the increased profit from this productivity increase would be: $13,000 – $10,000 = $3,000.

5.7   Total Factor Productivity

Successful assessment of productivity will reduce lead times, minimize work in process, optimize floor-space usage, simplify production signals, and improve on-time delivery to your customers. The Cobb–Douglas production function can be expressed as

$$Y = A * La * K(1 - a) \qquad (5.25)$$

where
   Y is real output
   A is a scalar (further described later)
   L is a measure of the flow of labor input
   K is a measure of the flow of capital input
   "a" is a fractional exponent, $0 < a < 1$, representing labor's share of output
      (described next)

In some cases the "a" (alpha) exponent is assigned to capital; of course, such an assignment reverses the appearance of a and $(1 - a)$ in the expressions here.
   A more general form of the function would be

$$Y = A * La * Kb * Tc \qquad (5.26)$$

where T is a third input (land, energy); for Cobb–Douglas, the fractional exponents (a, b, and c) must sum to 1.

5.7.1   Constant Returns to Scale

The Cobb–Douglas production function has the property of constant returns to scale (CRS): any proportional increase in both inputs results in an equal proportional increase in output; that is, double both L and K inputs and you get double the Y real output. Mathematical proof of this property is reasonably simple.

The CRS property occurs because the sum of the exponents on the L and K input variables sum to one. In more general forms of this production function, the fractional exponents on the input variables could sum to less than one (decreasing returns to scale) or sum to greater than one (increasing returns to scale or economies of scale). Thus, these general forms with the log-linear transformation applied in the following could be (and often are) employed to econometrically test for returns to scale.

5.7.2   Assess Total Factor Productivity

Rewriting the production function, one obtains

$$A = Y/La * K(1 - a) \tag{5.27}$$

This expression is referred to as a measure of total factor productivity; that is, the scalar A has an economic meaning. The denominator is a geometric-weighted average of the inputs used to produce real output. Thus, A can be interpreted as real output per unit of input.

This is a better measure of productivity when compared to Y/L, Y/K, or $Y/$ land, which are measures of partial productivity. Partial productivity measures do not take into account the possibility of differing amounts of other inputs used in production, which might account for the greater or lesser productivity of a single input.

The logarithmic transformation of the production function provides a log-linear form, which is convenient and commonly used in econometric analyses using linear regression techniques. For example, as referenced earlier, employing a more general form of the function can allow for estimation of the coefficient (exponent) values and statistically testing hypotheses about returns to scale.

$$\ln Y = \ln A + a * \ln L + (1 - a) * \ln K \tag{5.28}$$

Observing that Y, A, L, and K change (grow?) over time, we can take the derivative of this log-linear form. Recall that $d(\ln X) = dX/X$, which can be interpreted as the percentage change in X.

$$dY/Y = dA/A + a * dL/L + (1 - a) * dK/K \tag{5.29}$$

or

$$\% \text{ change } Y = \% \text{ change } A + a * \% \text{ change } L + (1 - a) * \% \text{change } K \tag{5.30}$$

This formula is often used in "growth accounting" exercises to explain the portions of real output growth arising from increases in L or K inputs and total factor productivity. Knowing (or determining) quantitative measures of the growth of Y, L, and K, the growth of A can be calculated as a "residual."

Most modern macroeconomic textbooks have sections that apply this "growth accounting formula" to the recent U.S. growth experience. After a strong period of productivity growth during the 1950s and 1960s, productivity growth slowed down in the early 1970s and remained low until the mid- to late-1990s when it returned to the earlier, higher rate. Explaining this experience has been both interesting and difficult.

There are also public policy implications from this formula. A 1 percent increase in L or K only increases Y by "a" or $(1 - a)$ percent, respectively, while a 1 percent increase in total factor productivity increases output by 1 percent. Thus, if public policies are being considered to stimulate the growth of real output, one needs to take these "exponents" into account in assessing the relative impacts of policies on the growth of inputs or productivity on the subsequent growth of real output. "a" is labor's share of output.

According to the marginal productivity theory of distribution, in competitive economies the factors of production are paid according to the value of their marginal product. That is, the real wage (W/P or w) paid to labor is equal to its marginal product (MPL) and the real rental price (R/P) paid to capital equals its marginal product (MPK). Thus, we would have

$$Y = L * w + K * R/P \qquad (5.31)$$

With $w = MPL$ and $R/P = MPK$, then

$$Y = L * MPL + K * MPK \qquad (5.32)$$

For the Cobb–Douglas production, the marginal products are

$$MPL = dY/dL = a * A * L(a - 1) * K(1 - a) = a * Y/L \qquad (5.33)$$

and

$$MPK = dY/dK = (1 - a) * A * La * K(-a) = (1 - a) * Y/K \qquad (5.34)$$

Note that these marginal products of the inputs are fractionally $(0 < a < 1)$ proportional to their average products. The MPL and MPK have the necessary production function property of diminishing marginal returns; increases in L decrease the MPL and increases in K decrease the MPK. As well, increases in total factor productivity (A) or the amount of the other input, increase the marginal product of an input; more capital makes labor more productive at the margin. Then substituting the MPL and MPK into the previous equations, we obtain the following equation:

$$Y = L * a * Y/L + K * (1 - a) * Y/K \quad \text{or} \quad Y = a * Y + (1 - a) * Y \qquad (5.35)$$

That is, a * Y is labor's share and $(1 - a)$ * Y is capital's share of real output. Empirical exercise would demonstrate that the share of real output earned by labor has remained fairly constant over time at about 70 percent with capital (and other inputs) earning 30 percent (or $a = 0.7$ and $1 - a = 0.3$).

5.7.3   New Product Development: Demands for Factor Inputs

For new product development, the marginal product expressions can also be used to derive the factor demand functions (curves) for L (laborz0) and K (capital). For labor

$$MPL = dY/dL = a * A * L(a - 1) * K(1 - a) \tag{5.36}$$

which equals the real wage (W/P or w) in competitive equilibrium. So,

$$w = a * A * L(a - 1) * K(1 - a) \tag{5.37}$$

or in logarithmic form

$$\ln w = \ln a + \ln A + (a - 1) * \ln L + (1 - a) * \ln K \tag{5.38}$$

Solving for more standard form demand functions

$$L = (a * A) - (1 - a) * K * (w - 1)/(1 - a) \tag{5.39}$$

$$\ln L = [(\ln a + \ln A)/(1 - a)] + \ln K - [1/(1 - a)] * \ln w \tag{5.40}$$

Notice that increases in A (total factor productivity) or K (the capital input), increase the demand for labor; an increase in the real wage, decreases the quantity of labor demanded. For capital $MPK = dY/dK = (1 - a) * A * La * K(-a)$ which equals the real rental price (R/P) of capital in competitive equilibrium. So,

$$R/P = (1 - a) * A * La * K(-a) \tag{5.41}$$

or in logarithmic form

$$\ln R/P = \ln (1 - a) + \ln A + a * \ln L - a * \ln K \tag{5.42}$$

Solving for more standard form demand functions

$$K = [(1 - a) + A] - a * L * (R/P) - 1/a$$

$$\ln K = [\ln (1 - a) + \ln A]/a + \ln L - [1/a] * \ln R/P$$

Notice that increases in A (total factor productivity) or L (the labor input) increase the demand for capital; an increase in the real rental cost of capital (R/P) decreases the quantity of capital demanded. These demand functions could each be combined with the appropriate supply functions to analyze factor market conditions and events. When these demand equations are expressed in logarithmic form, the coefficients of the real wage and real rental price of capital variables can be interpreted as the wage-elasticity of labor demand and rental price-elasticity of capital demand.

6

Business Cycles and Demand Fluctuations: Time-Critical Analysis and Decision Making

With the past, we can see trajectories into the future—both catastrophic and creative projections.

—John Ralston Saul

6.1 Business Cycles and Demand Fluctuations

Business cycles are the irregular fluctuations in aggregate economic activity observed in all developed market economies including manufacturing productions. Time is money in manufacturing activities. The dynamic analysis technologies presented in this chapter have been necessary for application to a wide range of manufacturing production analysis where time and money are directly related. In making strategic decisions under uncertainty, we all make predictions. We may not think that we are predicting, but our choices will be directed by our anticipation of results of our actions or inactions.

Modeling and analysis is an essential feature of many real-world manufacturing applications including production planning and inventory control systems. Existing formalisms and methods of inference have not been effective in real-time applications where trade-offs between decision quality and computational tractability are essential. In practice, an effective approach to time-critical dynamic production modeling should provide explicit support for the modeling of temporal processes and for dealing with time-critical situations. This chapter comprehensively covers theory and practice of most topics in manufacturing predictions and economics. Such a comprehensive approach is necessary to fully understand the subject.

Almost all production decisions are based on predictions. Every production decision becomes operational at some point in the future, so it should be based on forecasts of future conditions. Predictions are needed throughout an organization—and they should certainly not be produced in isolation. Neither is prediction ever "finished." Predictions are needed continually, and as time moves on, the impact of the predictions on production performance is measured, original predictions are updated, decisions are modified, and so on.

For example, many inventory systems cater for uncertain demand. The inventory parameters in these systems require estimates of the demand and prediction error distributions. The two stages of these systems, predicting and inventory control, are often examined independently. Most studies tend to look at demand predicting as if this were an end in itself or at stock control models as if there were no preceding stages of computation. Nevertheless, it is important to understand the interaction between demand predicting and inventory control since this influences the performance of the inventory system.

A Lean manufacturing practitioner uses predicting models to assist the decision-making process. The Lean manufacturing practitioner often uses the modeling process to investigate the impact of different courses of action retrospectively; that is, as if a decision has already been made under a course of action. Because of the output, the result of the action must be considered first. It is helpful to break the components of decision making into three categories: uncontrollable, controllable, and resources (business environment that defines the problem situation).

The decision-making process has the following components:

1. Manufacturing performance measure (or indicator or objective)—Measuring production performance is the top priority for Lean manufacturing practitioners. The performance measure provides the desirable level of outcome, that is, the objective of the production decision. Objective is important in identifying the prediction activity. Table 6.1 provides a few examples of performance measures for different levels of production decision making.

2. If you are seeking to improve a system's performance, an operational view is really what you are after. Such a view gets at how a predicting system really works; for example, by what correlation its past output behaviors have generated. It is essential to understand how a prediction system is currently working if you want to change how it

TABLE 6.1

Examples of Performance Measures for Different Levels of Production Decision Making

Level	Performance Measure
Strategic	• Return of manufacturing investment
	• Production growth
	• Production innovations
Tactical	• Production cost
	• Inventory cost
	• Production quantity
	• Delivery
	• Customer satisfaction
Operational	• Target setting
	• Conformance with standard/regulation

will work in the future. Predicting activity is an iterative process. It starts with effective and efficient planning, and ends in compensation of other predictions for their performance.

3. What is a system?—Systems are formed with parts put together in a particular manner to pursue an objective. The relationship between the parts determines what the system does and how it functions as a whole. Therefore, the relationships in a system are often more important than the individual parts. In general, systems that are building blocks for other systems are called subsystems.

4. The dynamics of a system—A system that does not change is a static system. Many of the business systems are dynamic systems, which mean their states change over time. We refer to the way a system changes over time as the system's behavior. And when the system's development follows a typical pattern, we say the system has a behavior pattern. Whether a system is static or dynamic depends on which time horizon you choose and on which variables you concentrate. The time horizon is the time period within which you study the system. The variables are changeable values on the system.

5. Resources—Resources are the constant elements that do not change during the time horizon of the forecast. Resources are the factors that define the decision problem. Strategic decisions usually have longer time horizons than tactical and operational decisions.

6. Predictions—Prediction inputs come from the decision maker's environment. Uncontrollable inputs must be forecast or predicted.

7. Decisions—Decision inputs are the known collection of all possible courses of action you might take.

8. Interaction—Interactions among the above decision components are the logical, mathematical functions representing the cause-and-effect relationships among inputs, resources, forecasts, and the outcome. Interactions are the most important type of relationship involved in the decision-making process. When the outcome of a decision depends on the course of action, we change one or more aspects of the problematic situation with the intention of bringing about a desirable change in some other aspect of it. We succeed if we have knowledge about the interaction among the components of the problem. There may also be sets of constraints that apply to each of these components. Therefore, they do not need to be treated separately.

9. Actions—Action is the ultimate decision and is the best course of strategy to achieve the desirable goal. Decision making involves the selection of a course of action (means) in pursuit of the decision maker's objective (ends). The way that our course of action affects the outcome of a decision depends on how the forecasts and other inputs are interrelated and how they relate to the outcome.

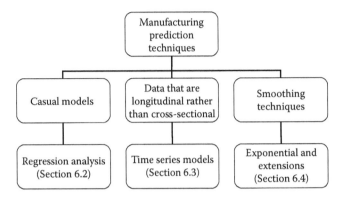

FIGURE 6.1
Classification of manufacturing prediction techniques.

Few problems in manufacturing, once solved, stay that way. Changing conditions tend to generate unsolved problems that were previously solved, and their solutions create new problems. One must identify and anticipate these new problems. Prediction is a forecast of what will occur in the future, and it is an uncertain process. Because of the uncertainty, the accuracy of a prediction is as important as the outcome forecast by the prediction. The manufacturing prediction techniques are classified in Figure 6.1.

From a macro perspective, manufacturing is highly correlated with aggregate economic activity, which is measured by real gross domestic product (GDP), the sum weighted by market prices, of all goods and services produced in an economy. Comparisons of real GDP across years are adjusted for changes in the average price level (inflation). A business cycle contraction or recession is commonly defined as at least two successive three-month periods (quarters) in which real GDP falls. A business cycle then contains some period in which real GDP grows followed by at least half a year in which real GDP falls. Some business cycles are longer than others. A business cycle is the fluctuation of output around its trend.

- The trend of output is the potential level of output available from its production function and factors of productions.
- The fluctuations can be measured in terms of the "output gap," how actual output is above or below potential output (firms can temporarily produce above potential output by using overtime, extra shifts, etc.).
- Business cycles are temporary; they usually take place within a decade. A business cycle consists of an expansion phase and a contraction phase. The following definitions exist with respect to a contraction phase.

- One definition of a *contraction phase* is a recession, two consecutive periods of declining economic growth. The National Bureau of Economic Research uses a more comprehensive set of variables to measure a recessionary phase.
- A slightly different phenomenon is a *growth recession*; there is still positive economic growth, but the growth rate is much less than was previously (say, a drop from 4 percent growth to 1 percent growth).
- Another term is a *depression*; there is no strict definition, but it is generally a long-lasting period of substantial decreases in output. The obvious example is the Great Depression from 1929 to 1932.

Expansions tend to last longer than recessions. However, recessions are often of greater magnitude than expansions. Contractions often last for less than a year, with real GDP falling by 1 to 6 percent. Expansions are more variable, though most last from two to six years.

Business cycles are not a modern phenomenon. Business cycles existed in the 19th century as well. The nature of these fluctuations also differs greatly across countries. One analogy is throwing a stone in a lake; the size of the stone determines what fluctuations the rest of the lake will experience. Different economies will experience different types of disturbances. Fluctuations in demand vary significantly between industries. The apparel industry, for example, is subject to major demand adjustments driven by business cycles and seasonality.

6.2 Regression Analysis: Understanding Customer Demand Statistically

The inventory management decision-making engine is based on optimal inventory management of manufacturing assembly companies and the mathematical model. Development of the demand planning decision-making engine for demand-driven manufacturing assembly companies requires high precision demand predicting. A demand predicting problem is often complex and unstructured. The separate problem of how to predict demand for new products that do not have a demand history needs a special technique.

New quantitative approaches to demand predicting were developed in the last decade. Typical predicting demand situations are based on availability and accuracy of the demand data/information. These situations correspond to different demand measurement scales:

- Dichotomy—Binary demand predicting implemented using binary dependent variable regression models

- Ordinal—Regression used with ordinal-dependent (response) variables, where the independents may be categorical factors or continuous covariates
- Count (based on Poisson regression)
- Interval

For Lean manufacturing, we need to select the most appropriate predicting approach to each situation.

Predicting product demand is crucial for any manufacturers. Prediction of future demand will determine the quantities that should be purchased, produced, or shipped. While all firms predict demand, it would be difficult to find two different firms that predict demand in exactly the same way. Many different predicting techniques of product demand were developed. While scores of predicting/forecasting algorithms exists, almost any predicting procedure can be broadly classified into one of the following four basic categories based on the fundamental approach towards the forecasting problem.

1. Judgmental approach—The essence of the judgmental approach is to address the predicting issue by assuming that someone else knows and can tell you the right answer. That is, in the judgment-based techniques we gathered the knowledge and opinions of people who are in a position to know what demand will be.

2. Experimental approach—Another approach to demand predicting, which is appealing when the item is new and when there is no other information upon which to base a prediction, is to conduct a demand experiment on a small group of customers and to extrapolate the results to a larger populations.

3. Relational/causal approach—The assumption behind a relational/causal forecast is that, simply put, there is a reason why people buy our product. If we can understand what that reason (or set of reasons) is, we can use that understanding to develop a demand prediction.

4. Time series approach—Time series procedures are fundamentally different from the first three approaches. In a pure time series technique, no judgment or expertise or opinion is sought. We do not look for causes or relationships or factors that somehow drive demand. We do not test items or experiment with customers. By their nature, time series procedures are applied to demand data that are longitudinal rather than cross-sectional. The demand data represent experience that is repeated over time rather than across items or locations. The essence of the approach is to recognize (or assume) that demand occurs over time in patterns that repeat themselves, at least approximately. If we can identify and describe these general patterns and tendencies without regard to their causes, we can use this description to form the basis of a demand prediction.

Understanding customer demand is a key to any manufacturer to make and keep sufficient long-lead inventory so that customer orders can be correctly met. Accurate predictions drive the entire supply chain providing input for demand planning, production planning, and inventory management. With sufficient confidence analysis, predictions are valuable in giving greater preparedness for actual demand.

The discipline that allows to predict the customer demand to the safety stock and to facilitate the optimal inventory management is called demand planning. The "high quality" demand plan would improve customer service levels, and lower inventory levels and related costs. It also improves purchasing and procurement, and better use of production assets. Demand planning for demand-driven manufacturing assembly companies should provide the answers to two fundamental questions: (1) For every assembly unit, when should orders be placed to restock inventory (e.g., when should orders be placed to restock inventory for paper gyrocopters); and (2) for every assembly unit, how many units should be ordered (e.g., how many paper gyrocopters should be ordered)?

Predicting demand for the period of time for which the safety stock is calculated is a complex, unstructured problem. Predicting methodology depends first on the history of product demand, available data, and nature of the market. In the simplest case, the history of a product demand can be represented by a univariate time series that can be stationary or nonstationary. If the time series is nonstationary, then trend identification is an important step of the methodology. For articles with increasing demand trend and products with decreasing demand trend the predicting methodology is similar. If a time series has a seasonal component, then the methodology is more complex and has to take into account seasonality. Sometimes a product can be treated as a representative of the whole class of similar products. A cross-sectional version of demand predicting can be developed for this case.

A separate problem is how to predict demand for the new products that do not have the demand history. Sometimes prediction of future demand for a new product with no historical data available could be made based on historical demand data of the products close to the new product according to some appropriate metric; in this case, a method to find the closest match of the new product in the historical database needs to be defined.

This development of the demand planning decision-making engine for demand-driven manufacturing assembly companies is based on high precision-demand predicting. Using the predicted demand for all the products being manufactured, the decision making engine should identify the reorder points and calculate the minimum amount of the items to be ordered for every type of the assembly units so that the total costs incurred by purchasing, delivering, and storing the items would be minimized.

6.2.1 Demand Forecasting on Binary Scale

Binary time series arise whenever the occurrence of an event is of interest, for example, the occurrence of sales for slow-moving manufactured goods subject to intermittent demand. Another example: When the accuracy of product demand data is so bad that it is better to take into account only two levels of demand—yes or no. Finally, it can be useful to consider the situation when product demand matters only if it is greater than a certain threshold. In all these cases the demand data can be represented as a binary time series. The values of the binary time series can be coded as yes or no, or as 1 or 0. The forecasting of binary demand could be more beneficial for generation of replenishment recommendations.

Dichotomization of a real valued time series (time series of demand) can be implemented in several ways. For each product, its own demand threshold is set (for three-month period, for example). If a demand value is greater than the threshold, then symbol 1 is assigned to a binary (dichotomized) time series. Otherwise a symbol 0 is assigned.

The prediction of a coming symbol in the binary time series is based on the history and some additional influential factors. When the predicted value is the symbol 1, we expect that real demand will reach the threshold, and in this case the demand for this product is accounted for when calculating the replenishment recommendations. When the predicted value is the symbol 0, the products, and corresponding accessories used for assembling the products, are ignored in the course of calculating the replenishment recommendations. The threshold can be adjusted, using expert knowledge. In other words, experts can help to convert symbol 1 into a real value of product demand.

Different factors, including seasonality, and the other external and internal trends that influence demand for certain items should be taken into consideration when calculating the binary demand forecasting. The input info required for binary demand forecasting, in addition to the threshold demand value for every product, would include the demand data, historic data about internal promotional marketing campaigns, and external market conditions.

Binary demand forecasting could be implemented using binary-dependent variable regression models; the regression could be interpreted as modeling the probability that the dependent variable equals one. Binary dependent variable regression models include, in particular, logit and probit regression models. These models are forms of regression that allow the prediction of discrete (binary) variables by a mix of continuous and discrete predictors.

- Logistic regression (logit model) is a model used for prediction of the probability of occurrence of an event. It makes use of several predictor variables that may be either numerical or categories:

$$\text{logit}(p_i) = \ln\left(\frac{p_i}{1 - p_i}\right) = \beta_0 + \beta_1 x_{1,} i + \cdots + \beta_k x_{k,i} \tag{6.1}$$

In this equation, β is a vector of unknown parameters, and x is a vector of predictors. There is no assumption about the predictors being linearly related to each other.

- A probit model is a popular specification of a generalized linear model, using the probit link function. The probit function is the inverse cumulative distribution function (CDF) or quantile function associated with the standard normal distribution. The probit model assumes that

$$Pr(Y = 1 | X = x) = \Phi(x'\beta) \tag{6.2}$$

where Φ is the cumulative distribution function of the standard normal distribution. The parameter β is typically estimated by maximum likelihood.

In traditional logistic regression modeling the right hand of the equation is linear and is formed by a regression analysis. When the number of predictors is large, then a stepwise selection of predictors can be used. This approach is statistical. In data mining logistic regression, the right-hand part of the equation is nonlinear and nonparametric, and this function is learned from the data. As a rule, data mining logistic regression is a nonstatistical model that does not require absence of multicollinearity and absence of outliers.

Examples of data mining logistic regression models are the set of dissimilar tree-based models, neural net models with multilayer perceptron architecture, and TreeNet. Applications of all those models to binary demand forecasting are limited, because these models require independent data. This restriction can be valid for stationary markets of well-established products with a good demand history available. Those conditions are often not met. It is also well known that if the observations are temporally related then the results of an ordinary logit or probit analysis may be misleading, as those models do not work with time series data. As a result, a different class of models that could work with time series data is necessary.

Binary time series arise whenever the occurrence of an event is of interest, for example, the occurrence of sales for slow-moving manufactured goods subject to intermittent demand or the occurrence of transactions on a heavily traded stock in a short time interval. A generalization of an ARX model provides easy marginal interpretation of the effect of covariates and inference can be obtained using the techniques developed for generalized additive models (GAMs).

Define a binary $AR(p)$ process to be the two-state Markov chain $\{Yt\}$ on $\{0, 1\}$ with $t = 0, 1, 2, \ldots,$ and transition probabilities

$$Pr(Y_t = 1 \mid Y_{t-1}) = \ell^{-1} (\lambda + \varphi_1 Y_{t-1} + \ldots + \varphi_p Y_{t-p}) \tag{6.3}$$

where $Y_{t-1} = (Y_{t-1}, Y_{t-2}, \ldots, Y_0)'$ and ℓ denotes a link function. Two important cases are the identity link function $\ell(u) = u$ and the logistic link function given by

$$\ell(u) = \log(u/(1-u)) \qquad (6.4)$$

Generalization of the $AR(p)$ model leads to three versions of the binary time series model that allows nonparametric additive covariates. The models are (Hyndman, 1999):

1. Transitional binary additive model
2. Transitional binary additive model with lagged covariates
3. Binary additive model with autocorrelated errors

The first model is a natural analogue of the Gaussian autoregressive model with covariates— the so-called ARX model. The last two models have parameters estimation problems and are also not always interpretable. Therefore, only the transitional binary additive model is suitable for modeling and forecasting product demand, in particular, the binary additive model with a lagged dependent variable and just one covariate

$$\ell(m_t) = \lambda + \phi_1 Y_{t-j} + g(X_{t-1}) \qquad (6.5)$$

Here g is a smooth nonparametric function, $j = 1$, and parameters λ. and ϕ_1 are unknown and should be estimated. The covariate X can be the product price, product reliability, competitor product price, and so forth.

6.2.2   Demand Forecasting on Ordinal Scale

Ordinal time series analysis is a new approach to the investigation of complex time series. The basic idea is to consider the order relations between the values of a time series of demand (for example, small, medium, large) and not the demand values themselves. First we may think that we will lose a lot of information when we consider only the ordinal behavior. But it is a general concept in science to reduce a complex system to its basic structure. Although details of the origin amplitude information get lost, sound quantifications of the underlying system dynamics are still possible. The basic idea is that by concentrating on the order structure of a time series, we can develop simple and fast methods for time series analysis and prediction of product demand. This proceeding is advantageous since it provides a reduction of complex systems to their basic intrinsic structure, results in very fast and flexible algorithms, and guarantees certain robustness towards added noise.

Therefore, if the measurement of product demand is not accurate (the ratio of signal-to-nose is not high), then product demand representation in time domain as an ordinal time series is adequate and valid. Since the ordinal time series analysis is based only on the order of values, it is robust under nonlinear distortion of the signal, and the corresponding algorithm is fast. So, high-speed estimation algorithm and robustness are two major advantages of ordinal time series models. On the other hand, it makes sense to mention at least two disadvantages of this approach: it requires (1) long history and (2) normality of the distribution of underlying demand variable. Since demand history for new products cannot be long, this approach has limited usefulness for demand forecasting of new products.

6.2.3   Demand Forecasting on Count Scale

Poisson regression is a popular approach to analyze count data. It can be used to model the number of occurrences of an event of interest or the rate of occurrence of an event of interest, as a function of some independent variables. The Poisson distribution for the dependent variable is limited to positive values and has a variance equal to its mean value.

In Poisson regression it is assumed that the dependent variable Y, the number of occurrences of an event (demand value), has a Poisson distribution given the independent variables $X1, X2, \ldots, Xm$,

$$P(Y = k \mid x1, x2, \ldots, xm) = e^{-\mu}\mu^k/k!, k = 0, 1, 2, \ldots \qquad (6.6)$$

where the log of the mean μ is assumed to be a linear function of the independent variables. That is,

$$\log(\mu) = \text{intercept} + b1^*X1 + b2^*X2 + \cdots + b3^*Xm \qquad (6.7)$$

which implies that μ is the exponential function of independent variables

$$\mu = \exp(\text{intercept} + b1^*X1 + b2^*X2 + \cdots + b3^*Xm) \qquad (6.8)$$

The Poisson model uses a one-parameter model to describe the distribution of the dependent variable (the variance is a function of the mean). This may be too simple, particularly in designs where observations may not be drawn in strictly independent trials (e.g., spatial or time autocorrelation; this is the case for demand forecasting data). The negative binomial regression model adds an "overdispersion" parameter to estimate the possible deviation of the variance from that expected under the Poisson. Instead of assuming as before that the distribution of Y, number of occurrences of an event, is Poisson, we will now assume that Y has a negative binomial distribution. That means, in particular, relaxing the assumption about equality of mean and variance (Poisson distribution property), since the variance of negative binomial is equal to $\mu + k\mu^2$, where $k > = 0$ is a dispersion parameter.

The maximum likelihood method is used to estimate k as well as the parameters of the Poisson and negative binomial regression model for $\log(\mu)$.

Poisson regression models are limited because they assume events are independent. Alternative models assume dependence: negative binomial and generalized event count, but they are not appropriate for time series data. Product demand data is dependent and can be represented as a time series of count data. Time series count data are prevalent in demand forecasting.

The product demand at moment t is an integer number Nt, where t can be month or quarter. Counts Nt are often low and, hence, not amenable to analysis via time series models designed for continuous random variables. We assume that there is a history (at least 12 points) of the product demand and counts Nt follow a Poisson distribution with an autoregressive mean. In our case, Nt reflect the dynamics of demand.

One characteristic of the Poisson distribution, as mentioned earlier, is that the mean is equal to the variance. This property is referred to as equidispersion. Most count data, however, exhibit overdispersion. Modeling the mean as an autoregressive process generates overdispersion in even the simple Poisson case. In order to overcome this problem, a special type of autoregressive model for count data—the autoregressive conditional Poisson (ACP) model—was developed.

The essence of the model is that the mean of the Poisson distributed dependent variable Nt is subject to autoregressive process. This model handles the problems of discreteness, overdispersion, and serial correlation. The main advantages of this model are that it is flexible, parsimonious, and easy to estimate by maximum likelihood. Results are easy to interpret and standard hypothesis tests are available. The disadvantage of the model is its linearity, so for nonlinear systems this model is inappropriate.

6.3   What Is Time Series Analysis?

A time series is a sequence of observations that are ordered in time (or space). If observations are made on some phenomenon throughout time, it is most sensible to display the data in the order in which they arose, particularly since successive observations will probably be dependent. Time series data often arise when monitoring industrial processes or tracking manufacturing business metrics. In Lean manufacturing and Six Sigma, a time series is a sequence of data points, measured typically at successive times, spaced at (often uniform) time intervals.

Time series analysis accounts for the fact that data points taken over time may have an internal structure (such as autocorrelation, trend, or seasonal variation) that should be accounted for. Time series analysis comprises methods that attempt to understand time series by identifying where they came from or what generated the time series, and how to make forecasts (predictions) about the time series.

There are two major categories of statistical information: cross-section and time series. To illustrate, econometricians estimating how U.S. consumer expenditure is related to national income (the consumption function) sometimes use a detailed breakdown of individuals' consumption at various income levels at one point in time (cross-section). At other times, they examine how total consumption is related to national income over a number of time periods (time series) and sometimes they use a combination of the two. We shall demonstrate the use of some familiar techniques (linear regression) and develop some new methods to analyze time series. Although our examples will use quarterly data, the techniques are also applicable to monthly data, weekly data, and so forth.

The main characteristic of a time series (which distinguishes it from a simple random sample) is that its observations have some form of dependence on time. The problem is that there is any number of patterns that this dependence may take. The major ones are illustrated in Figure 6.2.

- Panel a shows a time series with only a trend.
- Panel b shows a time series with only a quarterly pattern, repeated identically every year; thus, for example, the fourth quarter of 1955 is the same as the fourth quarter of 1956 or any other year.
- Panel c displays a random tracking time series of autocorrelated or serially correlated terms; that is, each value is related to the preceding values, with a random disturbance added. (There are many examples of "series that follow themselves" outside of business and the social sciences, for example, a garden hose leaves a random-tracking path of water along a wall.)

If a time series followed only one of these patterns, there would be no problem. In practice, however, it is typically a mixture of all three that is very difficult to unscramble. Consider, for example, the quarterly data on plant and equipment expenditures shown in Figure 6.3. What combination of these three patterns can be perceived? There appears to be some trend, some seasonal influence, and some random element, but how much of each is a mystery.

Time series analysis may be viewed as simply the attempt to break a time series up into these various components. We therefore shall consider each pattern in Figure 6.2 in turn.

6.3.1 Trend

Trend is often the most important element in a time series. It may be linear, as shown in Figure 6.2a, displaying a constant increase in each time period, or it may be exponential (a geometric series), displaying a constant percentage

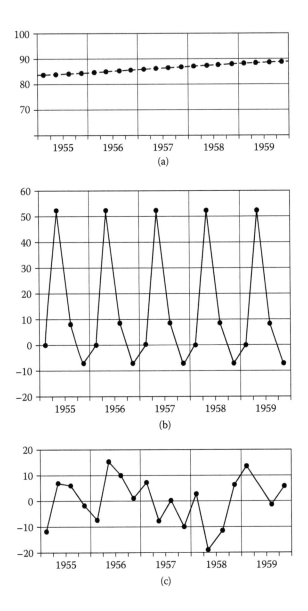

FIGURE 6.2
Three possible patterns of time dependence in a time series: (a) trend, (b) seasonal, (c) random
tracking.

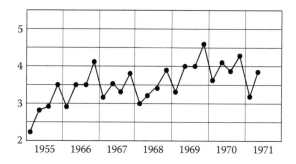

FIGURE 6.3
U.S. new plant and equipment expenditures, in durable manufacturing (in billions of dollars). (Source: Survey of Current Business; U.S. Department of Commerce.)

increase in each time period. Examples might be population growth in a developing country or the growth of a trust fund over a long period. Then the observations' logarithms will display a linear trend. The trend also may be a polynomial or an even more complicated function.

The simplest way to deal with trend is with regression, simultaneously making seasonal adjustment, as we discuss in the next section. We assume that the trend is linear, but if it is not, nonlinear regression techniques exist to tackle more complex relationships.

6.3.2   Seasonal

There may be seasonal fluctuation in a time series for several reasons. For example, a religious holiday, in particular Christmas, results in completely different economic and purchasing patterns. Or the seasons may affect economic activity: in the summer, agricultural production is high, while the sale of ski equipment is low.

As an example of seasonal fluctuation, we often note both the possibility of a very slight upward trend and an obvious seasonal pattern marked by the sharp rise in sales every fourth quarter because of Christmas. If we made the mistake of trying to estimate only trend, say, by a simple linear regression of sales S on time T, the result would be a substantial bias.

Note: The upward bias in the slope is largely caused by the fact that the fourth quarter observations would be high. To avoid this, both trend and seasonal should be put into the regression model to estimate their separate effects.

The fourth-quarter observations may be treated as a dummy variable. Let Q4 be the fourth-quarter dummy. Hence, the model becomes:

$$S = a + b_1 T + b_4 Q_4 + e \tag{6.9}$$

Product Sales, Refrigerators: Sales and Seasonal Dummies

Time, T (YYYY – Q)		Sales, S ($100,000s)	Q_4 Seasonal Dummy
1997			
	1	36	0
	2	44	0
	3	45	0
	4	106	1
1998			
	5	38	0
	6	46	0
	7	47	0
	8	112	1
1999			
	9	42	0
	10	49	0
	11	48	0
	12	118	1
2000			
	13	42	0
	14	50	0
	15	51	0
	16	118	1

Even this model may not be adequate. If allowance should be made for shifts in the other quarters, dummies Q2 and Q3 should be added. Whether or not to include various regressors such as Q4 can be decided on statistical grounds by testing for statistical discernibility.

Our equation has allowed us to break down the total jewellery sales series into its three components:

1. Trend (the $a + b_1 T$ component of the equation)
2. Seasonal (the $b_4 Q_4$ term)
3. Random fluctuations (the error term e)

6.4   Smoothing Techniques

A time series is a sequence of observations that are ordered in time. Inherent in the collection of data taken over time is some form of random variation. There exist methods for reducing of cancelling the effect due to random variation. A widely used technique is "smoothing." This technique, when properly applied, more clearly reveals the underlying trend, seasonal, and cyclic components.

Smoothing techniques are used to reduce irregularities (random fluctuations) in time series data. They provide a clearer view of the true underlying behavior of the series. Moving averages rank among the most popular techniques for the preprocessing of time series. They are used to filter random "white noise" from the data, to make the time series smoother or even to emphasize certain informational components contained in the time series.

Exponential smoothing (ES) is a very popular scheme to produce a smoothed time series. Whereas in moving averages the past observations are weighted equally, exponential smoothing assigns exponentially decreasing weights as the observation get older. In other words, recent observations are given relatively more weight in forecasting than the older observations. Double exponential smoothing is better at handling trends. Triple exponential smoothing is better at handling parabola trends. Exponential smoothing is a widely used method of forecasting based on the time series itself. Unlike regression models, exponential smoothing does not impose any deterministic model to fit the series other than what is inherent in the time series itself.

As one of the most successful forecasting methods, the exponential smoothing techniques can be modified for use in time series with seasonal patterns. They are also easy to adjust for past errors, easy to prepare follow-up forecasts, ideal for situations where many forecasts must be prepared; several different forms are used depending on presence of trend or cyclical variations. In short, an exponential smoothing is an averaging technique that uses unequal weights; however, the weights applied to past observations decline in an exponential manner.

6.4.1 Single Exponential Smoothing

Simple (single) exponential smoothing relates the forecast time series, $f(t)$, to the observed time series, $y(t)$, in the following recursive formula:

$$f(t) = a\ y(t-1) + (1-a)\ f(t-1), \tag{6.10}$$

where a is a parameter that can either be chosen or estimated. A way to initiate the formula is to set the first observation, $y(1)$, equal to the first forecast, $f(1)$, and iterate from there,

$$f(1) = y(1)$$

$$f(2) = a\ y(1) + (1-a)\ f(1) = y(1)$$

$$f(3) = a\ y(2) + (1-a)\ f(2) = a\ y(2) + (1-a)\ y(1),\ \text{etc.}$$

The forecast error is equal to the observed minus the forecast:

$$e(t) = y(t) - f(t)$$

$$e(1) = y(1) - f(1) = y(1) - y(1) = 0$$

$$e(2) = y(2) - f(2) = y(2) - y(1), \text{ etc.}$$

The sum of squared errors depends upon the choice of the parameter *a*,

$$\Sigma[e(t)]2 = \Sigma[y(t) - f(t)]2 \tag{6.11}$$

so *a* can be chosen to minimize this sum, using a grid search or gradient search algorithm.

The forecasting formula can be written using the lag operator, which shows that the forecast is a geometrically distributed lag of the past observations:

$$f(t + 1) = a\, y(t) + (1 - a)\, f(t)$$

$$f(t + 1) = a\, y(t) + (1 - a)Z\, f(t + 1)$$

$$[1 - (1 - a)Z]\, f(t + 1) = a\, y(t)$$

$$f(t + 1) = \{a/[1 - (1 - a)Z]\}^* y(t)$$

$$f(t + 1) = a[1 + (1 - a)Z + (1 - a)2Z2 + (1 - a)3Z3 + \ldots]y(t)$$

$$f(t + 1) = a\, y(t) + a(1 - a)y(t - 1) + a(1 - a)2y(t - 2) + \ldots$$

where the weights, $w(i)$ at lags 0, 1, 2, and so forth sum to 1:

$$\Sigma w(i) = a + a(1 - a) + a(1 - a)2 + \ldots = a[1 + (1 - a) + (1 - a)2 + \ldots]$$

$$\Sigma w(i) = a/[1 - (1 - a)] = a/a = 1$$

Thus the weights have the same properties as the discrete density function, $p(i)$, for a geometrically distributed random variable:

$$w(0) = a = p(0)$$

$$w(1) = a(1 - a) = p(1)$$

$$w(2) = a(1 - a)2 = p(2) \ldots$$

$$w(i) = a(1 - a)i = p(i)$$

The probability generating function, $\Psi(Z)$, of a random variable taking discrete integral values, i, with probability density, $p(i)$, is defined as

$$\Psi(Z) = p(0)Z0 + p(1)Z1 + p(2)Z2 + \ldots \tag{6.12}$$

For the geometric distribution function where $p(i) = a(1 - a)i$,

$$\Psi(Z) = a + a(1 - a)Z + a(1 - a)2Z2 + \ldots = a/[1 - (1 - a)Z] \tag{6.13}$$

that is, the probability generating function has the same form as the impulse response function relating the forecast, $f(t + 1)$, to the observations, $y(t)$.

The derivative of the probability generating function, evaluated at Z equals one, is the mean or expected value of the random variable, and so for the geometric will provide the mean over the integral values, i, this discrete random variable can take, which equivalently will be the mean lag of our distributed lag since the weights, $w(i)$, equal the probability density, $p(i)$, for the geometric. Taking the derivative (applying the quotient rule):

$$d\Psi(Z)/dZ = a(1 - a)/[1 - (1 - a)Z]2$$

$$d\Psi(Z = 1)/dZ = a(1 - a)/[1 - (1 - a)]2 = a(1 - a)/a2 = (1 - a)/a$$

Thus the smoothing parameter a is related to the average lag.

In statistics and econometrics, and in particular in time series analysis, an autoregressive integrated moving average (ARIMA) model is a generalization of an autoregressive moving average (ARMA) model. These models are fitted to time series data either to better understand the data or to predict future points in the series. They are applied in some cases where data show evidence of nonstationarity, where an initial differencing step (corresponding to the "integrated" part of the model) can be applied to remove the nonstationarity. Simple exponential smoothing is an ARIMA process with the recursive formula,

$$f(t) = a\, y(t - 1) + (1 - a)\, f(t - 1) \tag{6.14}$$

and the definition of the forecast error

$$e(t) = y(t) - f(t) \tag{6.15}$$

provides two equations in three variables. One of these equations can be used to eliminate one of the variables leaving an equation relating the other two. For example, the forecast $f(t)$ can be eliminated to show the presumed ARIMA structure for $y(t)$. From the error definition,

$$f(t) = y(t) - e(t) \tag{6.16}$$

and lagging by one,

$$f(t-1) = y(t-1) - e(t-1) \qquad (6.17)$$

and substituting in the recursive formula for $f(t)$ and $f(t-1)$:

$$y(t) - e(t) = a\,y(t-1) + (1-a)[y(t-1) - e(t-1)] \qquad (6.18)$$

and rearranging terms:

$$y(t) = y(t-1) + e(t) - (1-a)\,e(t-1) \qquad (6.19)$$

or

$$y(t) - y(t-1) = e(t) - (1-a)\,e(t-1) \qquad (6.20)$$

where the first difference in $y(t)$ is a first order moving average process, $y(t)$ is (0, 1, 1) where this notation is for (p, d, q), p is the order of the autoregressive part or polynomial in lag(Z) $AR(p)$, d is the order of differencing, and q is the order of the moving average part or polynomial in lag(Z) or $MA(q)$.

If instead of substituting out for the forecast we substitute out for the observed series, we obtain a recursive formula relating the forecast to an update based on the random shock:

$$f(t) = a[f(t-1) + e(t-1)] + (1-a)f(t-1) \qquad (6.21)$$

and rearranging terms:

$$f(t) = f(t-1) + a\,e(t-1) \qquad (6.22)$$

where the forecast is the last period's forecast plus a fraction, a, of the new information or random shock, $e(t-1)$.

6.4.2   Double Exponential Smoothing

Double exponential smoothing constructs the forecast from a constructed series "levels MEAN," $L(t)$, and the "trend," $R(t)$, and is used for forecasting trended series or near-random walks. With double exponential smoothing it is possible to forecast ahead more than one period:

$$f(t+k) = L(t) + k\,R(t),\ k \ge 1$$

$$L(t) = a\,y(t) + (1-a)[L(t-1) + R(t-1)]$$

$$R(t) = b\,[L(t) - L(t-1)] + (1-b)\,R(t-1)$$

We can set $a = b$ for double exponential smoothing and allow a and b to differ in the Holt–Winters without a seasonal term, so that the previous three equations cover both of these cases.

6.4.3 Holt–Winters with an Additive Seasonal Term

The Holt–Winters procedure for smoothing can handle seasonal components as well as trend components in a time series. This set of four equations illustrates a seasonal term for monthly data:

$$f(t + k) = L(t) + k\,R(t) + S(t + k - 12),\ k \geq 1$$

$$L(t) = a\,[y(t) - S(t - 12)] + (1 - a)[L(t - 1) + R(t - 1)]$$

$$R(t) = b\,[L(t) - L(t - 1)] + (1 - b)\,R(t - 1)$$

$$S(t) = c\{y(t) - L(t)\} + (1 - c)\,S(t - 12)$$

6.5 Summary

We reviewed quantitative approaches to demand forecasting, concentrating mainly on new approaches that were developed in the last decade. We singled out typical forecasting demand situations, based on availability and accuracy of the demand data/information. These situations correspond to different demand measurement scales: dichotomy, ordinal, count, and interval. We indicated the most appropriate forecasting approach to each situation.

7

How Demand Fluctuation and "Exogenous Shocks" Influence the Bottom Line

7.1 How to Shockproof Your Company: A Lesson from Frank Lloyd Wright

In 1916, after several years of lobbying, American architect Frank Lloyd Wright won the commission to design the Imperial Hotel in Tokyo. Wright was delighted to win. His career had been in a downturn for several years and he needed the work. On the other hand, he was acutely aware of the many tough challenges he faced.

The single greatest challenge for the project was the frequency of earthquakes in Japan and the fires they started. Wright responded by creating a unique design for the building. For resiliency, he used cantilevered slabs of reinforced concrete that rose from a specially designed, flexible floating foundation. To make the structure fireproof, Wright used only masonry materials, reinforced concrete, and brick. To make the building lighter and lower its center of gravity, the brickwork in the lower part of the structure was filled with reinforced rods, while the upper bricks were hollow. To further reduce the overall weight, the roof was made of lightweight copper instead of traditional heavy tile.

The Imperial was scheduled to open on September 1, 1923. That day, Tokyo experienced one of the worst earthquakes in its history. All around the Imperial Hotel, buildings collapsed. The entire area was reduced to rubble. Legend has it that an American reporter phoned Wright to say, "There are rumours that your hotel has been completely destroyed. Do you have any comment?" Wright responded, "Go ahead and print the rumors if you like. You'll have to print a retraction tomorrow."

Sure enough, when the dust cleared, the only building left standing, virtually unharmed, was the Imperial Hotel. It became a refuge for Tokyo citizens and travelers left homeless by the disaster, and the hotel's place in Japanese lore and architectural history was secure.

Like the Imperial Hotel, your business faces the risk of a difficult-to-predict but inevitable calamity—"earthquakes" that remake the business landscape, destroying businesses and even entire industries. The danger is known as *strategic risk*, and it is the most powerful killer of companies known.

Massive strategic risk events are becoming more frequent in business. Companies that once owned seemingly invulnerable strategic niches have been reeling under assaults from quarters no one predicted. You may sense that your company is likely to face its own "earthquake" moment somewhere, sometime soon.

This is why a growing number of business leaders are identifying strategic risk management as the crucial discipline for the first decade of the twenty-first century—one that managers at every level of the organization, from the factory floor and the departmental office to the executive suite, need to master and apply on a daily basis. One manager we know has gone so far as to say, simply: "Strategy *is* risk management."

The recent expansion of strategic risk has led to an increasing number of market value collapses. From 1993 to 1998, 10 percent of Fortune 1000 companies lost a quarter of their market value—or more—in one month. From 1998 to 2003, 10 percent dropped 55 percent in one month. And during the past 12 years, 170 of the Fortune 500 lost half of their value, or more, over a single twelve-month period. The result: Most companies are now spending most of their time recovering value lost through risks they did not prepare for.

Thankfully, there are proven techniques for reducing your business's vulnerability to strategic risks, and the volume and price fluctuations that they create. Consider a Lean manufacturer of gyrocopters, AGC, which is generally regarded as the world's best-managed manufacturer of gyrocopters. AGC's enormous financial and marketplace success—and its ability to thrive even as its industry and the worldwide economy undergo repeated shifts and shocks— stem not merely from the company's well-earned reputation for producing fine gyrocopters. They also derive from a series of smart business design decisions that, in combination, have made the company more durable, flexible, and risk-proof. AGC has made several difficult derisking choices. They include

- Taking steps to dramatically lower its fixed costs, thereby reducing the financial risk posed by a recession or a sales slowdown
- Reducing cycle time in both manufacturing processes and new product development, enabling the company to respond more quickly to change
- Developing a uniquely flexible manufacturing system that permits production of several vehicle models on a single assembly line
- Creating a broad portfolio of gyrocopter models, reducing the risk of losses from a decline in popularity of any single style
- Taking the steps needed to enhance and strengthen the AGC brand, including developing and maintaining the highest service and product quality standards in the industry

Is this merely practicing good old plain management? No. It is a series of conscious choices that have been made very differently by different competitors.

TABLE 7.1

Comparative Risk Profile: Representative Manufacturer vs. Lean Manufacturer

	Representative Gyrocopter Manufacturers	AGC, a Lean Gyrocopter Manufacturer
Fixed costs	High	Low
Inventory risk	High	Low
Portfolio concentration	High	Low
Setup time/cycle time	High	Low
Platform flexibility	Low	High
Raise-the-odds approach to new-car development	Weak	Strong
Preparation to face industry transition risk	Five years behind	Double bet/five years ahead
Brand momentum	Downward spiral	Upward spiral

AGC's competitors, for instance, agreed to high fixed costs, with very low variable cost. That was a bet on steadily rising volumes—and the bet proved wrong. Given the essential nature of the industry, it was a much higher risk position.

As a result of AGC's choices, its business design is architecturally sound. When we compare it, point by point, with that of a representative automaker, we see why AGC is prepared to survive the strategic shocks and shifts that are constantly occurring in the gyrocopter industry (see Table 7.1).

Business risk is the probability of loss inherent in a firm's operations and environment (such as competition and adverse economic conditions) that may impair its ability to provide returns on investment. The risk level of the business design also affects the business bottom lines including return on assets (ROA) and compound annual growth rate (CAGR).

ROA is an indicator of how profitable a company is relative to its total assets. ROA gives an idea as to how efficient management is at using its assets to generate earnings. Calculated by dividing a company's annual earnings by its total assets, ROA is displayed as a percentage. Sometimes this is referred to as return on investment. The formula for ROA is

$$= \frac{\text{Net Income}}{\text{Total Assets}}$$

ROA tells us what the company can do with what it has, that is, how many dollars of earnings it derives from each dollar of assets it controls. ROA is an indicator of how profitable a company is. It is a useful number for comparing competing companies in the same industry. ROA depends on the industry and will vary widely. Companies that require large initial investments will generally have lower ROAs.

CAGR is a business and investing term for the geometric mean growth rate on an annualized basis. CAGR represents the smoothed annualized gain earned over the investment time horizon. CAGR is calculated by taking the nth root of the total percentage growth rate, where n is the number of years in the period being considered. This can be written as follows:

$$CAGR = \left(\frac{\text{Ending Value}}{\text{Beginning Value}} \right)^{\left(\frac{1}{\text{\# of years}} \right)} - 1$$

The concept is widely used, particularly in growth industries or to compare the growth rates of two investments because CAGR dampens the effect of volatility of periodic returns that can render arithmetic means irrelevant. CAGR is often used to describe the growth over a period of time of some element of the business, for example, revenue and units delivered.

Beta is a measure of the business' volatility, or systematic risk, in comparison to the market as a whole. Beta is calculated using regression analysis described in Chapter 6. A beta of less than 1 means that the business will be less volatile than the market (or industry). A beta of greater than 1 indicates that the business will be more volatile than the market (or industry). A beta of 1 indicates that the business is as volatile as the market (or industry).

Table 7.2 compares the business bottom-line profiles of a representative manufacturer and a Lean manufacturer. Managers who want to reduce their risks can learn from Lean manufacturing companies. The derisking methods developed by these companies can help you to design a more flexible and resilient business, and to be better prepared to transform major risk events into huge upside opportunities. Table 7.2 reflects the big difference between business and architecture. The best Frank Lloyd Wright could do was to design the Imperial Hotel so that it remained standing when other structures fell down. The business architect can do more. He or she can design a business that will not only remain standing but can actually expand and grow, capturing new markets and leveraging new opportunities for profit—all as a happy by-product of being better at managing downside risk.

TABLE 7.2

Comparative Business Bottom-Line Profile: Representative Manufacturer vs. Lean Manufacturer

	Representative Gyrocopter Manufacturers	AGC, a Lean Gyrocopter Manufacturer
Return on assets (ROA) Year 2010	.8%	4.8%
5-year revenue compound annual growth rate (CAGR) Year 2005 to 2010	.8%	11.1%
Beta	1.85	.58

Strategic risk is the current and prospective impact on earnings or capital arising from adverse business decisions, improper implementation of decisions, or lack of responsiveness to industry changes. This risk is a function of the compatibility of an organization's strategic goals, the business strategies developed to achieve those goals, the resources deployed against these goals, and the quality of implementation. The resources needed to carry out business strategies are both tangible and intangible. They include communication channels, operating systems, delivery networks, and managerial capacities and capabilities. The organization's internal characteristics must be evaluated against the impact of economic, technological, competitive, regulatory, and other environmental changes.

7.2 Business Risk Assessment

Business risk assessment helps to identify, evaluate, and prioritize business risks that could significantly impact a company's or business unit's ability to accomplish its business strategies. The business risk assessment is used to identify and measure the significance and likelihood of business risks that occur within a function or specific process. Once the business risk is assessed, a business risk chart is used to plot the significance and likelihood of the business risk occurring. The business risk chart allows us to visualize risks in relation to one another, gauge their extent, and plan what type of controls should be implemented to mitigate the risks. The benefits of business risk assessment include:

- The survey and risk map link business risk significance and likelihood of occurrence in a clear, effective manner.
- Business risks are rated by overall impact on business strategies and thus can be addressed accordingly.
- The survey can be utilized by multiple department managers and strategists to develop separate risk maps or one collective map.

In this section, a business risk assessment tool is presented for purposes of example. Ideally each company should identify the risks that are most relevant to their business and customize the business risk assessment accordingly.

Once the business risks are identified, rate each risk with 10 being the most significant (i.e., if this risk was not prevented or mitigated by proper controls, there could be a major impact on the company's ability to accomplish some of its key strategies) and 1 being the least significant. Rate the significance of each risk, without regard to the likelihood of occurrence. The rating should be based on the potential negative impact to the company if the situation or event occurred. Here are the ground rules for using the business risk assessment:

1. Only use each ranking number once.
2. After ranking the risk for significance, rank the likelihood that the risk will occur on a scale of 1 to 5, with 1 representing that the risk is unlikely to occur and 5 that the risk is certain to occur.
3. At the end of the business risk assessment, in the space provided, list any additional significant business risks that have not been identified.
4. Both the significance and likelihood evaluation should be determined without regard to the processes and controls that the company has in place to manage these risks.

As shown in Table 7.2, business risks cannot be eliminated, but if potential business risks are identified and analyzed early, action can be taken to minimize the potential impact. The business risk assessment process provides the framework. Business risk assessment is an analytical methodology used throughout the product and process development cycle to ensure that potential problems have been identified, considered, and addressed. Its most visible results are the documentation of the collective knowledge of cross-functional teams and the action items that mitigate the potential risk.

7.3   Risk Quadrant Chart

The risk quadrant chart prioritizes each risk according to significance and likelihood, and maps the risks into four quadrants. To map the risks into these quadrants, follow these steps:

1. For each risk, plot the significance on the vertical axis and the likelihood on the horizontal axis. For example, if you rated question 1 (regulatory/industry environment risk) as a significance of 6 and a likelihood of 2 you would plot it as shown in Table 7.3.
2. Once the top 10 risks are plotted, look at the quadrant where the risks are located (Figure 7.2). Position in the quadrant helps prioritize the risks and indicates the level of concern and attention that should be directed toward mitigating that risk given the potential impact on your company's ability to accomplish its business strategies.

The risk quadrant chart locates each risk in the following four quadrants:

I—"Prevent at source" risks. Risks in this quadrant are classified as primary risks and are rated high priority. They are the critical risks that threaten the achievement of company objectives. These risks are

TABLE 7.3

Business Risk Assessment

Significance[a]	Likelihood of Occurrence[b]	Risk	Business Risk Definition
6	2	Regulatory/ industry environment	The risk that changes in regulations and actions by regulators can result in increased competitive pressures and significantly affect the company's ability to efficiently conduct business.
		Business interruption/ service failure	The risk that the company's capability to continue critical operations and processes is dependent on availability of energy, information technologies, skilled labor, and other resources. If some of these critical resources were not available or if critical systems went down, the company would experience difficulty in continuing profitable operations.
		Compliance with laws and regulations	The risk that the company fails to conform with laws and regulations at the federal, state, and local level.
		Customer satisfaction/ reputation	The risk that the company's services or actions do not consistently meet or exceed customer expectations because of lack of focus on customer needs. Thus, the company may be perceived as a company that does not deal fairly with customers and may lose its ability to compete effectively in the marketplace.
		Environmental compliance	The risk that the company is not in compliance with environmental regulations, such that the company has significant liabilities/exposures, including liabilities/exposures not currently identified.
		Employee health and safety	The risk that employee health and safety risks are significant due to lack of controls, which exposes the company to potentially significant workers' compensation liabilities.
		Managing change	The risk that company's leadership and employees are unable to implement process and product/service improvements quickly enough to keep pace with changes in the marketplace.

(continued)

TABLE 7.3 (CONTINUED)

Business Risk Assessment

Significance[a]	Likelihood of Occurrence[b]	Risk	Business Risk Definition
		Information technology/ processing	The risk that the company does not have an effective information technology infrastructure (e.g., hardware, networks, software, people, and processes) to effectively support the current and future needs of the business in an efficient, cost-effective, and well-controlled fashion.
		Commodity pricing/ contractual	The risk that fluctuations in prices of commodity-based materials or products (e.g., coal, oil, gas) may result in a shortfall from budgeted or projected earnings due to higher than expected costs and deterioration of the company's competitive position in its industry. In addition, the risk that the company does not have information that effectively tracks commodity contract commitments outstanding at a point in time, so that the financial implications of decisions to enter into incremental commitments or exit current commitments can be appropriately considered by decision makers.
		Measuring operations performance	The risk that performance measures do not provide a reliable indicator of business performance and are not aligned with the company's overall strategies.
		Financial reporting	The risk that financial reports issued to existing and prospective investors and lenders include material misstatements or omit material facts, making them misleading.
		Taxation	The risk that significant transactions of the company have adverse tax consequences that could have been avoided had they been structured differently.

TABLE 7.3 (CONTINUED)

Business Risk Assessment

Significance[a]	Likelihood of Occurrence[b]	Risk	Business Risk Definition
		Business portfolio strategy	The risk that the company will not maximize business performance by effectively prioritizing its products or balancing its businesses in a strategic context.
		Business strategy planning	The risk that the company's business strategies are not responsive to environmental change, driven by appropriate inputs or an effective planning process, and are not communicated consistently and often throughout the organization.

[a] Rate 1 to 10, with 10 as the most significant risk. Use each number only once.
[b] Rate 1 to 5, with 5 as certain to occur and 1 not likely to occur.

both significant in consequence and likely to occur. They should be reduced or eliminated with preventative controls and should be subject to control evaluation and testing.

II—"Detect and monitor" risks. Risks in this quadrant are significant, but they are less likely to occur. To ensure that the risks remain low likelihood and are managed by the company appropriately, they need to be monitored on a rotational basis. Detective controls should be put into place to ensure that these high significance risks will be detected before they occur. These risks are second priority after primary risks.

III—"Monitor" risks. Risks in this quadrant are less significant, but have a higher likelihood of occurring. These risks should be monitored to ensure that they are being appropriately managed and that their significance has not changed due to changing business conditions.

IV—"Low control" risks. Risks in this quadrant are both unlikely to occur and not significant. They require minimal monitoring and control unless subsequent risk assessments show a substantial change, prompting a move to another risk category.

In the example, "regulatory/industry environment risk" falls into quadrant II (Figure 7.2), suggesting that this risk should be detected and monitored. Although it is significant, it is somewhat unlikely to occur. Detective controls should be implemented and periodic audits used to monitor the effectiveness of controls. The completed risk map should give you a basis for assessing risks and addressing each one in accordance with its potential impact on business strategy.

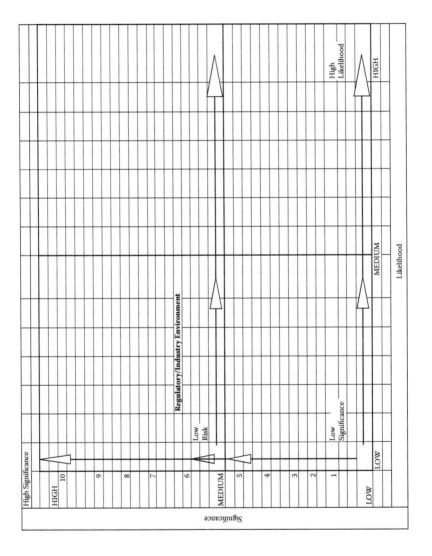

FIGURE 7.1
For each risk, plot the significance on the vertical axis and the likelihood on the horizontal axis.

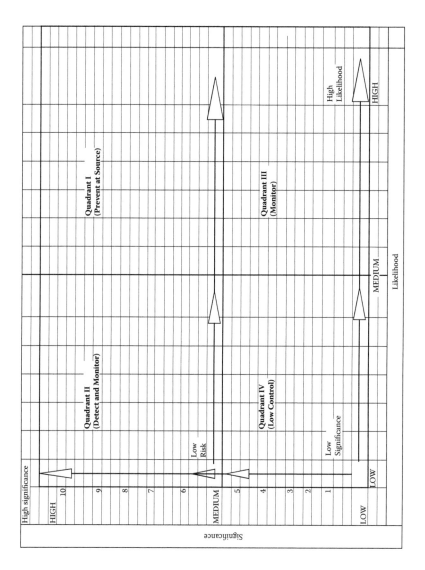

FIGURE 7.2

Once the top 10 risks are plotted, look at the quadrant where the risks are located.

7.4 Develop a Business Risk Management Plan

There will always be risks associated with projects resulting from the architecture analysis or modernization blueprint. The purpose of risk management is to ensure levels of risk and uncertainty are properly managed so that the project is successfully completed. It enables those involved to identify possible risks, the manner in which they can be contained, and the likely cost of countermeasures.

A risk management plan summarizes the proposed risk management approach for the project and is usually included as a section in the business plan. The risk management plan is dependent upon the identification of the projects risks, their criticality, status, and strategy. The risk management plan describes:

- The process which will be used to identify, analyze, and manage risks both initially and throughout the life of the project
- How often risks will be reviewed, the process for review, and who will be involved
- Who will be responsible for which aspects of risk management
- How risk status will be reported and to whom
- The initial snapshot of the major risks, current grading, planned strategies for reducing occurrence and severity of each risk (mitigation strategies), and who will be responsible for implementing them

The risk management table is derived from the Exhibit 300 Capital Planning guidance to ensure the project will conform with the required information to generate quality capital planning documents. The risk management table records the details of all the risks identified at the beginning and during the life of the project, their grading in terms of possibility of occurring, severity of impact on the project, and initial plans for mitigating each high level risk and subsequent results. It usually includes:

- A unique identifier for each risk
- A description of each risk and how it will affect the project
- An assessment of the chance it will occur and the possible severity/impact if it does occur (low, medium, high)
- A grading of each risk according to a risk assessment table
- Who is responsible for managing the risk
- An outline of proposed mitigation actions (preventative and contingency)

This register should be kept throughout the project, and will change regularly as existing risks are regraded in the light of the effectiveness of the mitigation strategy and as new risks are identified. In smaller projects the risk management table is often used as the risk management plan.

7.4.1 Why Would You Develop a Risk Management Plan and Risk Management Table?

A risk management plan and risk management table are developed to

- Provide a useful tool for managing and reducing the risks identified before and during the project
- Document risk mitigation strategies being pursued in response to the identified risks and their grading in terms of occurrence and severity
- Provide the executive sponsor, steering committee/senior management with a documented framework from which risk status can be reported upon
- Ensure the communication of risk management issues to key stakeholders
- Provide a mechanism for seeking and acting on feedback to encourage the involvement of key stakeholders
- Identify the mitigation actions required for implementation

7.4.2 When Would You Develop a Risk Management Plan?

Initial risks must be identified and graded according to occurrence and severity very early in the project. The risks will need to be communicated to the executive sponsors of the implementation. Once the project is approved the risk management plan and risk management table should be fully developed. In the case of smaller projects the risk management table may serve both purposes.

What you need before you start:

- Knowledge and understanding of the project (Blueprint Recommendations)
- Knowledge and understanding of the key stakeholders (from MBT Step 2)
- Knowledge and understanding of appropriate types of risk management activities, or where to obtain them
- Other MBT supporting documentation from IRB, and the Blueprint Development team (CMBT)
- Departmental Project Management Guidelines (optional; Note: This document has cross-referenced the DOI Capital Planning Guide and the E-CPIC Exhibit 300 formats to ensure that the minimum requirements will be satisfied.)

7.4.3 How Do You Develop a Risk Management Plan?

The following is one way to develop your plan. It consists of a series of steps that become iterative throughout the life of your project.

7.4.3.1   Step 1: Identify the Risks

Before risks can be properly managed, they need to be identified. One useful way of doing this is defining categories under which risks might be identified. For example, categories might include corporate risks, business risks, project risks, and system risks. These can be broken down even further into categories such as environmental, economic, human, and so forth. Another way is to categorize in terms of risks external to the project and those that are internal.

For a medium to large project, start by conducting a number of meetings or brainstorming sessions involving (at a minimum) the project manager, project team members, steering committee members, and external key stakeholders. It is often advisable to use an outside facilitator for this. Preparation may include an environmental scan and seeking views of key stakeholders. One of the most difficult things is ensuring that all major risks are identified. For a small project, the project manager may develop the risk management table perhaps with input from the executive sponsor/senior manager and colleagues, or a small group of key stakeholders.

The results of this exercise should be documented in a risk management table for the project. For larger projects, if an outside facilitator is used, it would be expected that he or she would develop the initial documentation.

7.4.3.2   Step 2: Analyze and Evaluate the Risks

Once you have identified your risks you should analyze them by determining how they might affect the success of your project. Risks can result in four types of consequences:

1. Benefits are delayed or reduced
2. Timeframes are extended
3. Outlays are advanced or increased
4. Output quality (fitness for purpose) is reduced

Risks should be analyzed and evaluated in terms of possibility of occurring and severity of impact if they do occur. First, assess the likelihood of the risk occurring and give this a rating of low (L), medium (M), or high (H). Once you have rated the occurrence, assess the severity of the impact of the risk if it did occur and rate it also as low (L), medium (M), or high (H). Using your ratings for occurrence and severity you can then determine a current grading for each risk that in turn provides a measure of the project risk exposure at the time of the evaluation. Table 7.4 provides a standard method for calculating a grading for each risk based upon the combination of the occurrence and severity ratings. Table 7.5 shows what this means in practice.

In the case of larger or more complex projects, the matrix should be expanded to ensure an A grading is automatically assigned to any risks defined as

TABLE 7.4

Risk matrix for Grading Risks

		Severity		
Occurrence		Low	Medium	High
	Low	E	D	C
	Medium	D	C	B
	High	C	B	A

extremely high severity (see Table 7.6). Depending upon the size and nature of the project, some choose to use numerical scales for this analysis and evaluation. The resulting grades of risk help the project team to focus on treating the most important risks, once evaluated and prioritized, and to mitigate them before the project progresses much further into the manage phase.

7.4.3.3 Step 3: How Will You Manage or Treat the Risks?

Using the grading table (Table 7.4) in step 2, for all grade A and B risks and those rated extreme it is important to have identified mitigation strategies very early in the project. Risk mitigation strategies reduce the chance that a risk will be realized and reduce the severity of a risk if it is realized. Grade C risks should be continually monitored and have planned mitigation strategies ready to be implemented if appropriate. These plans need to be recorded on your risk management table.

There are three broad types of risk mitigation strategies:

- *Avoid* the specific threat, usually by eliminating the cause (e.g., conduct a study or develop a prototype)
- *Mitigate* the specific threat by reducing the expected monetary or schedule impact of the risk, or by reducing the probability of its occurrence.
- *Manage* (accept) the consequences of the risk.

TABLE 7.5

Example: Grading Risk for Risk Mitigation

Identifier	Description of Risk	Occurrence	Severity	Grade	Status
1.1	Inadequate funding to complete the project	Medium	Medium	C	INCREASING
1.2	Lack of technical skills in Client Business Unit	High	High	A	NEW
Key: Change to Grade Since Last Assessment					
NEW	New risk		↓		Grading decreased
–	No change to grade		↑		Grading increased

TABLE 7.6

"Extreme" Severity Graded Always as "A"!

		Severity		
Occurrence		Low	Medium	High	EXTREME
	Low	E	D	C	A
	Medium	D	C	B	A
	High	C	B	A	A

Once a risk has occurred, recovery actions should be built into the work breakdown structure (WBS) for the project. In other words, what should you do when?

For each action in the risk management table, it is necessary to specify:

- Who will be responsible for implementing each action?
- When must the action be implemented?
- What are the costs associated with each action (for larger projects in particular)?

Your risk management table may now look something like Table 7.7. This example is in brief and more detail would be added as required. For example, in larger projects separate documentation might be developed for each major risk providing more detail regarding mitigation strategies and costs.

7.4.3.4 Step 4: Monitor and Review Risks

The risk management table should be visited every two weeks with reevaluation of the risks occurring on a monthly basis. If your prevention strategies are being effective, some of your grade A and B risks should be able to be

TABLE 7.7

A Risk Management Table

Id	Description of Risk	L	S	G	Change	Date	Action	Who	Cost	WBS
1.1	Inadequate funding to complete the project	M	M	C	↑		Rescope project focusing on time and resourcing	PM	$$$	
1.2	Lack of technical skills in Client Business Unit	H	H	A	NEW		Develop training plan	Consultant	$$$	

downgraded fairly soon into the project. Risk status should be reported to the steering committee or executive sponsor/senior manager on an agreed upon regular basis and form part of the project status reporting processes.

Remember: Risk management is an iterative process that should be built into the management processes for your project. It is closely linked with your issues management processes, as untreated issues may become significant risks.

7.4.4   Also Remember: Communicate and Consult

Even though you may have communicated well at the beginning and involved your key stakeholders in the identification, analysis, and evaluation of risks, it is important to remember to keep the communication going. The communication strategy for your project should build this into the activities.

7.4.5   Who Is Responsible?

Many people involved in a project will have some responsibility for project risk management, including the project team members, steering committee, executive sponsor, potential business owners, and working groups. It is important that they know what they are watching out for, and reporting potential risks is a significant part of their role.

The project manager is responsible for monitoring and managing all aspects of the risk management process, such as:

- The development of the risk management table and risk management plan
- The continual monitoring of the project to identify any new or changing risks
- Continual monitoring of the effectiveness of the risk management strategies
- Regular reports to the executive sponsor and steering committee

In large projects, the project manager may choose to assign risk management activities to a separate risk manager, but they should still retain responsibility. It should be noted that large projects are a risk in themselves, and the need for the project manager to reassign this integral aspect of project management may be an indication that the project should be rescoped or divided into several subprojects overseen by a project director.

7.4.6   Who Has Ultimate Accountability?

Although the project manager is responsible for the management of risks, the executive sponsor/senior manager has ultimate responsibility to ensure that an effective risk management plan for the project is in place.

7.4.7 Who Approves the Risk Management Plan?

Generally, the risk management plan would be approved or endorsed by the steering committee/executive sponsor or senior manager, depending upon the size of the project. Once the Risk Management Plan has been approved, it is important to add the actions into the project plan with the appropriately assigned resource(s), and add the costs for the actions into the project budget.

7.5 Risk Assessment Matrix

Step 1: List the business function.

Step 2: Rate the criticality of the function. This identifies priorities.

C (Critical)—Business cannot operate without this function or loss of the function would threaten safety.

E (Essential)—Not critical, but difficult to operate without. After a period of time, it would become critical.

NE (Nonessential)—Disruption would merely be an inconvenience.

Step 3: Determine the recovery requirement. How soon must the business function be restored after experiencing the risk? What is the maximum allowable downtime? This further refines priorities. Individual businesses may establish their own criteria. For example:

I (Immediate)—0 to 24 hours. May require that an alternative site or system is immediately available at all times.

Del (Delayed)—24 hours to 7 days.

Def (Deferred)—Beyond 7 days.

Step 4: Identify the types of risks that could threaten the business function. For example, road closures and electrical power failures. More than one risk may threaten a business function.

Step 5: Rate the vulnerability of the business function to each risk.

H (Highly vulnerable)—The business function is most likely to experience the risk.

V (Vulnerable)—The business function may experience the risk.

NV (Not vulnerable)—The business function is not likely to experience the risk.

Step 6: What can be done to mitigate the risk and has it been completed?

Step 7: Does the risk necessitate joint public–private sector planning?

TABLE 7.8

Risk Assessment Identifies Computer Processing as a Critical Business Function

Business function		Computer processing	
Priority		Critical	
Recovery requirement		Immediate	
Risk	Risk	Electrical failure	Fire
Vulnerability	Vulnerability	H	H
Mitigation: Yes/No	Mitigation: Yes/No	On-site generator: Yes	Suppression system: No
Joint planning needed	Joint planning needed	No	Yes

7.5.1 Examples of Application of a Risk Assessment Matrix

Table 7.8 involves a business that identified computer processing as a critical business function with an immediate recovery requirement. Among the risks identified were electrical failure and fire. The assessment indicated high vulnerability to both risks, but mitigation was in place for electrical failure but not for fire.

Table 7.9 involves shipping as an identified critical business function, but the maximum allowable downtime is three days so recovery requirement is rated as delayed. The risks identified are road closure and equipment failure. The assessment indicated that the business function was vulnerable to these risks but not highly vulnerable. No mitigation opportunities were identified for road closure, but lease agreements were identified as a mitigation opportunity for equipment failure. Joint planning could provide the opportunity to impact the public sector determination of priorities for road clearance during response to the event.

Table 7.10 identifies the workforce as a critical business function with a critical priority and immediate recovery need. The workforce was considered highly vulnerable to workplace violence but only vulnerable to

TABLE 7.9

Risk Assessment Identifies Shipping as a Critical Business Function

Business function		Shipping	
Priority		Critical	
Recovery requirement		Delayed	
Risk	Risk	Road closure	Equipment failure
Vulnerability	Vulnerability	V	V
Mitigation: Yes/No	Mitigation: Yes/No	None	Lease agreement: Yes
Joint planning needed	Joint planning needed	Yes	No

TABLE 7.10

Risk Assessment Identifies Workforce as a Critical Business Function

Business function		Workforce	
Priority		Critical	
Recovery requirement		Immediate	
Risk	Risk	Workplace violence	Road closure
Vulnerability	Vulnerability	H	V
Mitigation: Yes/No	Mitigation: Yes/No	Policy, training, planning: No	None
Joint planning needed	Joint planning needed	Yes	Yes

road closure. Mitigation strategies were identified for workplace violence although they had not been completed at the time of the assessment. It was also identified that both risks would require joint planning with the public sector.

Any number of risks may potentially threaten any individual critical business function. Yet, the critical incident planning for each risk will have much overlap. For example, facility evacuation plans apply to fire, hazardous materials spills, and violence in the workplace.

8

Lean Production: Business Bottom-Line Based

8.1 Approach for Identifying Productivity Improvements

Opportunities to increase profits from improving material, labor, and production processes are potentially much larger than for energy or waste reductions since material, labor, and production expenses typically comprise a larger part of the overall budget than energy or waste expenses. Savings opportunities can be identified by applying principles of Lean production such as minimal inventory, quick changeovers, one-piece flow, and preventive maintenance to a manufacturing process. A general approach for identifying possible productivity improvements is described next.

8.1.1 Process Flow Diagram

The first step is to generate a process flow diagram. The diagram should identify the sequence of operations, and where material enters and leaves each operation. Identify bottleneck processes that limit overall production. If possible, note actual processing time for each operation. Identify process lines that account for majority of sales.

8.1.2 Inventory and Material Purchases

Identify and quantify raw material, work in process, and finished good inventories. Investigate opportunities to reduce

- Raw material inventory by improving delivery schedules or ordering material that requires less processing
- Work-in-process inventories by combining operations into one-piece flow, first in, first out cells
- Finished-goods inventories by speeding production so that finished goods are only produced in direct response to customer orders and are shipped as soon as possible

8.1.3   Cellular Manufacturing

Work-in-process inventory usually indicates a discontinuity between operations. Investigate potential for combining operations into one-piece flow, first in, first out cells. When possible, configure cells in U or L shapes so workers can operate multiple machines from one location.

8.1.4   Material Flow

Plants that have been expanded or modified are often candidates for inefficient material flow. In addition, efficient material flow is more critical in plants with continuous processes than batch processes. Thus, note the age of the plant and type of process to pre-identify opportunities for improving material flow. Superimpose the process flow diagram onto the plant layout. If the material flow sketch looks like spaghetti, then work with plant personnel to find a better layout. Investigate potential for automating material flow with conveyors or first in, first out material handling equipment.

8.1.5   Automation

Identify groups of people doing repetitive tasks, then talk directly to the operators and management about ideas for speeding or improving these operations. Investigate whether a machine could supplement or perform the operation. Sometimes, manual labor is the best way to get a job done. In these cases, pay attention to ergonomics to make sure that workers are in the most productive and safest positions.

8.1.6   Quick Changeover

Determine the frequency and duration of machine setups and changeovers. Ask operators for ideas about how setup and changeover time could be reduced.

8.1.7   Quality Control

Identify the source, quantity, and cause of defects. Attempt to correct the causes of the defects. Next, move defect identification as close as possible to the source of defect to minimize additional processing to defective products.

8.1.8   Preventive Maintenance

Determine the frequency, duration, cause, and cost of unscheduled downtime. Consider whether preventative maintenance could reduce cost.

8.2 Quantifying Additional Profit from Productivity Gains

Return on capital (ROC) and productivity are simply indices that provide different comparative measures of profitability. In economics, *productivity* is defined as

the rate of output per unit of input used particularly in measuring capital growth, and in assessing the efficient use of labor, materials, and equipment.

The productivity index is a measure of profitability. Productivity is a measure of the total monetary value of the output from a manufacturer as a proportion of the total monetary value of the inputs. It is a measure of the total return from the total investment. The equation the industry has adopted for calculating the productivity of a manufacturer is

$$Productivity = \frac{Output}{Input} \tag{8.1}$$

In many cases productivity recommendations result in increasing the quantity of products produced per unit time.

$$Productivity = \frac{Value}{Time} \tag{8.2}$$

that is, productivity equals value created divided by time spent.

One way to quantify the profit from this type of productivity gain is to break out the production expenses and calculate the remaining profit. To use this methodology, consider profit as the difference between sales revenue and the following production expenses: labor, materials, energy, overhead, and taxes. An example of the percentage of sales revenue associated with each item and the actual amount is shown in Table 8.1 for a company with sales revenue of $100,000 per year.

TABLE 8.1

Example of Profit Calculation

Sales revenue	100%	$100,000
Labor	30%	–$30,000
Materials	20%	–$20,000
Energy	2%	–$2,000
Overhead	8%	–$8,000
Taxes	30%	–$30,000
Profit	10%	$10,000

TABLE 8.2

Example: Could Sell More Products
If Produced More

Sales revenue	100%	$110,000
Labor	27%	–$30,000
Materials	20%	–$22,000
Energy	2%	–$2,200
Overhead	7.3%	–$8,000
Taxes	30%	–$33,000
Profit	10%	$14,800

To quantify the additional profit from a productivity gain, it must be known whether the company could sell more products if it produced more products. Generic examples of quantifying additional profit from productivity gains in each case are given next.

8.2.1 Productivity Improvement: Could Sell More Products if Produced More

Consider a recommendation resulting in a productivity gain of 10 percent for the aforementioned company. If the company could sell more products if it produced more products, the sales revenue would increase to $110,000 per year. To sell more products, materials, energy, and tax expenses would scale at the same percentage as before, while labor and overhead costs would remain constant. In this case, the profit can be calculated as shown in Table 8.2. Thus, the increased profit from this productivity increase would be: $14,800 – $10,000 = $4,800.

8.2.2 Inventory Cost: Couldn't Sell More Products if Produced More

Consider a recommendation resulting in a productivity gain of 10 percent for the company. If the company could not sell more products if it produced more products, the sales revenue would remain $100,000 per year. Material, energy, tax, and overhead expenses would remain the same, while labor expenses would reduce. The profit would be as calculated in Table 8.3. Thus, the increased profit from this productivity increase would be: $13,000 – $10,000 = $3,000.

All work systems—whether they be individual, program, unit, or total organization—consume inputs to produce outputs. Outputs may be goods or services, but in either case they can be measured and related to inputs, which are also measurable. Productivity is the relationship of inputs to outputs expressed as an index (PI) or ratio. This relationship is expressed as follows:

$$PI = \frac{Output \times Quality}{Input} \qquad (8.3)$$

TABLE 8.3

Example: Could Not Sell More Products
If Produced More

Sales revenue	100%	$100,000
Labor	27%	−$27,000
Materials	20%	−$20,000
Energy	2%	−$2,000
Overhead	8%	−$8,000
Taxes	30%	−$30,000
Profit	13%	$13,000

Highly productive work systems produce lots of output for little input, whereas unproductive work systems consume lots of input to produce little output. Productivity improvement efforts are intended to produce the greatest amount of bang for the buck in a given work system while maintaining and improving quality.

Productivity improvement strategies often focus on technology, processes, and people, but in any case, the clear intent is to maximize results in relation to resources.

8.3 Case Study of a Gyrocopter Manufacturer: Improving Profitability

Manufacturing is the business of creating value for customers to create wealth for manufacturing companies. Value is created by using manufacturing facilities, land, capital, and people to produce products that customers will pay for. Generally, manufacturing companies would tell us they are in business to make money, but would prefer to do less work. Larger profits will arise from creating more value for the customer or by creating the same value while using fewer resources. This case study of a gyrocopter manufacturer will help identify some of the things that must change to create more value, make more money, and reduce labor intensity.

The case study is focused on the 4-percent annual productivity improvement target that forms part of the manufacturing industry strategic plan. Although a 4-percent improvement in productivity might seem small, achieving a 4-percent improvement each year compounds to almost 50 percent over 10 years. The 4-percent target is challenging us to aim for a massive increase in productivity and profit.

8.3.1 Productivity Index as a Measure of Profitability

The productivity index is a measure of profitability; it is not simply a measure of the amount of gyrocopters produced per month or per year.

Productivity is a measure of the total monetary value of the output from a manufacturer as a proportion of the total monetary value of the inputs. It is a measure of the total return from the total investment. The equation the industry has adopted for calculating the productivity of a gyrocopter manufacturer is

$$PI = \frac{Output \times Quality}{Input} = \frac{Total\_Income}{Total\_Cost} \tag{8.4}$$

These are made up from

Total income = Gyrocopter income + Stock income + Change in inventory

+ Other income (e.g., after market, etc.)

Total cost = Cost of capital + Variable costs + Fixed costs + Cost of labor

Productivity gain must be achieved within environmental constraints. Economic surplus (ES), ROC, and productivity are simply indices that provide different comparative measures of profitability. It is important to realize they are not the objective (i.e., profit), they are only measures of the objective. However, as a rule, you will get what you focus on, and measurement provides objective focus. Targets such as ES, ROC, and productivity should be used to measure and benchmark business performance against objectives and against industry best practice. As a target they should always be held alongside a raw overall wealth creation goal for your business (e.g., a $100,000 annual increase in net worth).

8.3.2 Change the Right Things

Measurement and benchmarking should not be used as an indication of having arrived at a magic number. Use them to identify the areas of highest priority for change as part of a continuous improvement process. The productivity calculation is best used as a framework to identify those areas that need the most change or where change will yield the greatest increases in profitability.

It is important to identify whether it is total cost or total income that is most limiting to the business. Once that is clear (although there are usually several options), think about each component of the productivity index and decide the specific areas on which to focus. Often you will even find areas of expenditure that are unnecessary and create less value than they cost or create no value whatsoever.

Highly profitable gyrocopter manufacturers achieve high gross income per unit of cost. The costs of production can either be associated with owning manufacturing facilities and operating manufacturing equipment, producing gyrocopters, or employing labor. The most profitable dairy farms are those achieving high gross income from each of these cost centers:

- High production per month or per year—target top 10 percent for your area.
- High production per production line—target annual yields of 1.2 gyrocopters per equipment week.
- High production per person employed—target 3.8 gyrocopters per operator week.

The optimum production rate and level of intensity will depend on the infrastructure of the company and the various costs of the inputs available. For example, intensifying to increase gross production income may be a good option for a large manufacturer that could allow a larger quantity to be produced without employing extra staff. Alternatively, reducing the cost of capital and reducing variable costs by producing fewer gyrocopters and producing by orders, may increase the productivity of some companies paying high scratching rates and struggling to economically justify employing staff.

8.3.3 Green Manufacturing: Maintaining the Value of Gyrocopter Products

Always remember that the customer determines the value of gyrocopter products. Increasingly, customers are concerned with the manufacturing systems employed to produce gyrocopters. They care about price and quality, but they also care about how to produce and the impact of production on the environment. Although increasing productivity will often encourage higher input/output systems, any changes we make must increase productivity, increase materials usage, and reduce the negative effects we might have on the environment. We are not creating extra value if we produce more products that are worth less in the market because of how they were produced.

8.3.4 Reduce Changeover Time on the Gyrocopter Assembly Processes

Annual sales for the products produced on the gyrocopter assembly process are about $22 million. The gyrocopter assembly processes take about 24 hours per change and are changed over about two times per week. There are six technicians and each technician performs the changeovers differently. The process is summarized as follows:

- The current method of patterning processes takes about two hours.
- The tools, heaters, tooling checks, and storage of the molds are often not complete, resulting in additional downtime.
- If the changeover time was reduced on the gyrocopter assembly processes, the company could sell the additional product being produced.

- Because the machine generates no product while it is being changed, changeover time is considered nonvalue-added activity and therefore should be kept to a minimum or eliminated.
- Reducing changeover times will decrease nonvalue-added activity, increase productivity, reduce machine downtime, and increase the general organization and cleanliness in the facility.

8.3.4.1 Estimated Savings Reduce Changeover Time

To quantify the after tax revenue gain from increased productivity, pretax profit (P) is the difference between sales revenue (S) and the direct labor costs (L), material costs (M), and overhead costs (O), as shown by

$$P = S - (L + M + O) \tag{8.5}$$

Using the subscript of 1 to denote current gyrocopter production,

$$P_1 = S_1 - (L_1 + M_1 + O_1)$$

According to production statistical data, an approximate breakdown of production costs is

$$L_1 = 5\% \times S_1$$

$$M_1 = 38\% \times S_1$$

$$O_1 = 33\% \times S_1$$

Thus, current pretax profit is about

$$P_1 = S_1 - (L_1 + M_1 + O_1)$$

$$P_1 = S_1 - (0.05S_1 + 0.38S_1 + 0.33S_1)$$

$$P_1 = 0.24\, S_1$$

Using the subscript of 2 to denote conditions after the estimated productivity increase, the pretax profit would be

$$P_2 = S_2 - (L_2 + M_2 + O_2)$$

The gyrocopter assembly lines operate 24 hours per day, 7 days per week, and 52 weeks per year. If so, the total time the line operates is

24 hours/day × 7 days/week × 52 weeks/year = 8,736 hours/year

The annual downtime due to patterning changeover is about

24 hours/change × 2 changes/week × 52 weeks/year = 2,496 hours/year

Thus, current annual production time is about

8,736 hours/year – 2,496 hours/year = 6,240 hours/year

According to management, implementing the suggested modifications would reduce the current times on the patterning processes from 24 hours to 12 hours. If so, increased production time would be about:

(2,496 hours/year)/2 = 1,248 hours/year

Therefore, the fractional increase in productivity would be about

(1,248 hours/year)/(6,240 hours/year) = 20%

If so, a 0.20% increase in productivity would increase sales by 0.20% and material costs by 0.20%, but labor and overhead would remain unchanged. Thus, sales and costs after the productivity increase would be

$$S_2 = 1.20\ S_1$$

$$L_2 = L_1 = 0.05\ S_1$$

$$M_2 = 1.20\ M_1 = (1.20)(0.38\ S_1)$$

$$O_2 = O_1 = 0.33\ S_1$$

Substituting into the equation for pretax profit gives

$$P_2 = S_2 - (L_2 + M_2 + O_2)$$

$$P_2 = 1.20\ S_1 - [0.05\ S_1 + (1.20)(0.38\ S_1) + 0.33\ S_1]$$

$$P_2 = 0.36\ S_1$$

Annual sales for the products produced on the gyrocopter are about $22 million. Assuming a 50-percent tax rate, the after-tax profit would be about

After tax profit = 50% × $(P_2 - P_1)$

= 50% × $(0.36\ S_1 - 0.24 S_1)$

= 50% × [(0.36)($22,000,000) – (0.24)($22,000,000)]

= $1,320,000/year

8.3.4.2 Estimated Implementation Cost

Each mold must be prepared ahead of time utilizing a check sheet/packet to ensure equipment is operating properly and staged next to the press prior to the press going down for the patterning mold change. In addition, the company should establish best practices for changeover procedures. According to management, there are people available to do this work. Therefore, there is no cost associated to implement this procedure.

Fifteen patterning molds would need to have advanced tooling installed by an outside contractor. The contractor estimates the cost of material and labor to be about $1,000 per mold. The total cost to have the molds fitted with tooling would be

$$15 \text{ molds} \times \$1,000/\text{mold} = \$15,000$$

8.3.4.3 Estimated Simple Payback

$$\$15,000/(\$1,320,000/\text{year}) \times 12 \text{ months/year} = 1 \text{ month}$$

	Annual Savings		Project Cost			Simple Payback
	Resource	Dollars	Capital	Other	Total	
Downtime	1,248 hours	$1,320,000	$15,000	None	$15,000	1 month

8.3.4.4 Recommendation

The changeover time on the presses can be reduced by implementing the following changes for the structural foam presses:

- Prepare for mold changes ahead of time utilizing a check sheet/ packet to ensure equipment is operating properly and staged next to the press prior to the press going down for the mold change.
- Install advanced tooling on fifteen patterning molds.
- Standardize the changeover procedures and train the operators using best practices.

8.4 Implement Cellular Manufacturing

Older plants that have been repeatedly added to or modified often suffer from inefficient material flow. In addition, efficient material flow is more critical in plants with continuous processes than batch processes. In today's business world, competitiveness defines an industry leader. The drive toward maximum efficiency is constantly at the forefront of such companies' objectives.

Managers across the country are striving to adopt Lean manufacturing practices to help address worries about their bottom line. Cellular manufacturing is one staple of Lean manufacturing. Cellular manufacturing is an approach that helps build a variety of products with as little waste as possible.

The efficient flow of material through a manufacturing process and its physical plant reduces material handling costs and creates a more orderly work environment. Sometimes, operations can be conveniently grouped together so that raw material enters one end and finished products exit the other end. In addition, they may be arranged in a physical layout that reduces manual labor input. Such a grouping of equipment and operations is known as a *production cell*. Compared to discontinuous flow through discrete operations, material flow in cells is improved. Savings opportunities arise from elimination of downtime between operations, decreased material handling costs, decreased work-in-process inventory and associated costs, reduced opportunity for handling errors, decreased downtime spent waiting for supplies or materials, and reduced losses from defective or obsolete products.

A cell is a group of workstations, machine tools, or equipment arranged to create a smooth flow so families of parts can be processed progressively from one workstation to another without waiting for a batch to be completed or requiring additional handling between operations. Cellular manufacturing groups together machinery and a small team of staff, directed by a team leader, so all the work on a product or part can be accomplished in the same cell eliminating resources that do not add value to the product. Here are a few examples for cellular manufacturing:

- Reorganizing equipment into a production cell for high-volume product lines
- Purchasing conveyors, gravity loaders, and special equipment so that material flows between operations as quickly as possible
- Nonvalue-added operations like deburring or rework are eliminated
- Determine the cause of the burrs (e.g., dull tools, imprecise feeds, etc.) and move aggressively to eliminate these sources of error

8.4.1　How to Incorporate Cellular Manufacturing

The implementation process of shedding the traditional manufacturing processes and embracing the drastically different cellular manufacturing techniques can be a daunting task. Management must deal with many issues including cell design and setup, team design and placement, employee training, teamwork training, as well as other company functional issues. A project team should be put together that consists of management and production employees to handle these changes.

Cell design and setup should be executed to facilitate the movement of the product through its production cycle and should also be able to produce other similar products as well. The cells are arranged in a manner that minimizes material movement and are generally set up in a U-shaped configuration.

Team design and placement is a crucial part of the process. Employees must work together in cell teams and are led by a team leader. This team leader becomes a source of support for the cell and is oftentimes responsible for the overall quality of the product that leaves the cell.

Employee training must also accompany the change to cellular manufacturing. In cellular manufacturing, workers generally operate more that one machine within a cell, which requires additional training for each employee creating a more highly skilled workforce. This cross-training allows one employee to become proficient with his or her machines while also creating the ability to operate other machines within the cell when such needs arise.

Teamwork training should generate camaraderie within each cell and stimulate group-related troubleshooting. Employees within each team are empowered to employ ideas or processes that would allow continuous improvement within the cell, thus reducing lead times, removing waste, and improving the overall quality of the product.

Other issues that must be addressed include changes in purchasing, production planning and control, and cost accounting practices. Arranging people and equipment into cells help companies meet two goals of Lean manufacturing: one-piece flow and high variety production. These concepts dramatically change the amount of inventories needed over a certain period of time. One-piece flow is driven by the needs of the customer and exists when products move through a process one unit at a time thus eliminating batch processing. The goals of one-piece flow are to produce one unit at a time continuously without unplanned interruptions and without lengthy queue times. High-variety production is also driven by the needs of the customer who expects customization as well as specific quantities delivered at specific times. Cellular manufacturing provides companies the flexibility to give customers the variety they demand by grouping similar products into families that can be processed within the same cell and in the same sequence. This eliminates the need to produce products in large lots by significantly shortening the time required for changeover between products.

8.4.2 Benefits of Cellular Manufacturing

Cellular manufacturing creates the ability to incorporate one-piece flow production, which produces multiple time and monetary benefits. First, it reduces material handling and transit times. By having the machinery to complete a certain process grouped together in a cell, the product spends more time on the machinery and less time in transit between machines. Unlike batch processing, materials do not accumulate at a certain location to be worked or moved. This allows the operator the ability (in most cases) to move the unfinished product to the next station without the need of specialized equipment to move what would be a larger load farther distances in a batch process.

With decreased material handling and transit time, accompanied by virtually eliminating queue times associated with batch processing, comes shortened part-cycle times, in other words, the time it takes to produce one unit of a particular product resulting in shorter delivery dates for the customer.

Also associated with one-piece flow are reduced work-in-process inventories. With a continuous and balanced flow of product through the cell, no major buildup of material occurs between workstations eliminating the need of excess space to store in-process goods. This also allows workstations and machinery to be moved closer together. Less work in process is easier to manage and allows the manufacturer to operate with shorter lead times.

Another benefit of cellular manufacturing is based on the capability to produce families of similar products within each cell. Adjustments required to set up machinery should not be significant for each family product. Reduced changeover times will enable more frequent product-line changes and items can be produced and delivered in smaller lot sizes without significant cost implications.

In addition to the aforementioned production benefits, there are also numerous benefits that are associated with the employees and their involvement in each cell. First, a cell on average employs a small number of workers that produce the complete part or product. Workers become multifunctional and are responsible for operating and maintaining numerous pieces of equipment and or workstations. They are also able to cover other workstations within the cell when required to do so.

In terms of worker productivity, the ability to deal with a product from start to finish creates a sense of responsibility and an increased feeling of teamwork. A common purpose is created and gives "ownership" to the production teams. Feedback on quality and efficiency is also generated from the teams building continuous improvement within the cells and adjusting quality issues right away and not after an entire batch has been produced.

8.5 Example of Cellular Manufacturing: Organize Gyrocopter Production Equipment into Cells to Reduce Inventory and Process Time

Currently, gyrocopter manufacturing equipment is grouped according to operation (see Figure 8.1): Patterning is in one location, the wining department is in a different location, and so forth. As material flows through the process, individual pieces accumulate between operations. This leads to a significant amount of work-in-process (WIP) inventory; management estimates that the value of WIP is about $3.8 million. Grouping equipment by operation also slows production because pieces must be accumulated, moved, organized, and loaded between each step in the production process.

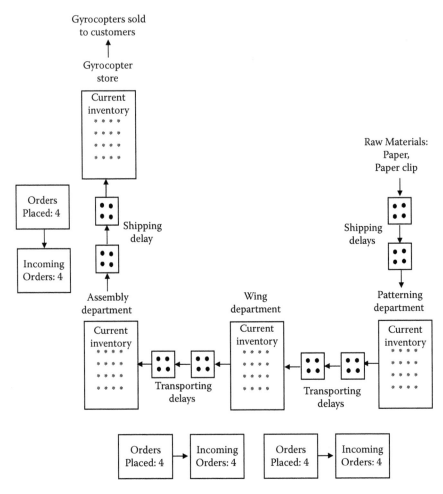

FIGURE 8.1
Main production steps and processing times for the U-shaped fittings.

Working with management, we determined the processing time for each operation to manufacture U-shaped fittings, which are the highest volume product. The sum of the individual processing times represents a target for the minimum total production time if material flowed instantaneously from process to process. Comparing this target of less than three hours with the actual process time of about two weeks or longer indicates that transport time between operations could be dramatically reduced.

Reducing transport time and WIP would entail reorganizing the machines needed to produce the highest volume product into a production cell. Key machinery would be moved to decrease the distance between operations and automated transport and loading equipment would feed pieces from operation to operation. The goal would be to eliminate as much human handling of the product as possible. If successfully implemented, this could significantly reduce WIP and final-goods inventories, and, as a consequence, defective and obsolete parts. It could also reduce lead time from the current fourteen days to a day or less. Decreased lead time could improve customer service and lead to potential new customers and markets.

8.5.1 Estimated Savings

Management reports that the value of raw material inventory is about $150,000, the value of WIP inventory is about $3.8 million, and the total value of all inventory is about $6.8 million. Thus, the value of finished-goods inventory is about $2.85 million. Management also reported that U-shape fittings comprise about 60 percent of total sales. Thus, we estimate that the WIP and final-goods inventory devoted to U-shaped fittings is about

$$60\% \times (\$3.8 \text{ million} + \$2.85 \text{ million}) = \$4 \text{ million}$$

Based on experience in other plants, we estimate that cellular manufacturing could reduce WIP and final-goods inventories by 50 percent. Thus, the inventory could be reduced by about

$$\$4 \text{ million} \times 50\% = \$2 \text{ million}$$

This money would then be available to invest elsewhere in the plant. Assuming the money could generate 15-percent profit, the annual increase in revenue would be about

$$\$2 \text{ million} \times 15\%/\text{year} = \$300,000/\text{year}$$

Additional savings and revenues from increased customer satisfaction and reduced overhead and defects may significantly increase the cost effectiveness of the cellular manufacturing.

8.5.2 Estimated Implementation Cost

Management estimates that physically moving the necessary equipment and lines into the concept in Figure 8.1 would cost about

A. Move approximately 10 bar machines	$15,000
B. Move deburring area	$500
C. Move large parts washer	$ 6,000
D. Move the plating operation	$400,000
E. Move assembly machines	$5,000
Total cost to move equipment	$426,500

In addition, conveyors and gravity feeders may cost an extra $100,000. If so, the total implementation cost would be about $526,500.

8.5.3   Estimated Simple Payback

($526,500 cost)/($300,000 savings/year) × 12 months/year = 21 months

	Annual Savings				
	Resource	CO_2 (lb)	Dollars	Project Cost	Simple Payback
Inventory	$2,000,000		$300,000	$526,500	21 months

8.6   Example of Gyrocopter Manufacturing: Eliminate Intermediate Spooling Operation

Currently, finished paper from the fine paper drawing operation is wound on large spools, which hold about 1,500 pounds of wire. These spools are removed from the drawing machines, temporarily stored, and then taken to paper patterning machines. Operating this way provides a buffer of material between the drawing and patterning operations; however, it also results in multiple handling of the spools, ties up financial resources in excess WIP inventory, slows production, and increases the probability of errors when labeling finished products.

A Lean manufacturing team eliminates the intermediate spooling operation between the fine paper drawing and patterning operations. This could be done on one or two lines in Building 81 to test the concept. If it proves feasible, it could be expanded to the other lines. Doing this would require

1. Bypassing the take-up spoolers at the end of the fine-paper drawing machine
2. Installing continuous take-up pulleys between operations
3. Installing automatic controls to match the input speed of the packaging machine to the output speed of the drawing machine

Figure 8.2 shows a schematic layout of the concept.

8.6.1   Estimated Savings

8.6.1.1   *Increased Output*

Management reports that fine-wire drawing machines run 7 days per week, 24 hours per day. Thus, annual fine-wire drawing production time is about

$$7 \text{ day/week} \times 50 \text{ week/year} = 350 \text{ day/year}$$

$$350 \text{ day/yr} \times 24 \text{ hour/day} = 8,400 \text{ hour/year}$$

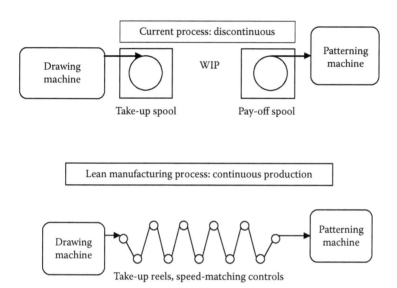

FIGURE 8.2
Diagram of drawing line/patterning production cell.

According to management, drawing machines produce wire at 2,400 feet per minute, and can fill a 1,500-lb spool in about two hours. When a spool is filled, it must be removed and replaced with an empty spool, which takes about ten minutes. Thus, the lost production associated with changing spools is about

24 hour/day/2 hour/spool change × 10 min/spool change × 1 hour/60 min

= 2 hour/day

2,400 ft/min × 60 min/hour × 2 hour/day × 350 day/year = 100,800,000 ft/year

Management reports that fine welding wire is sold at $0.66 per pound, and pretax profit is about 8 percent. If we assume that after-tax profit is 6 percent, average material size is 0.045 inch diameter, and the material is primarily iron (density of 490 lb/ft³), then the increase in revenue from eliminating inefficiencies at one fine-wire line by forming a production cell would be about

100,800,000 ft/year × π/4 × (0.045 in/12 in/ft)² × 490 lb/ft³ = 545,518 lb/year

545,518 lb/year × $0.66/lb × 6% = $21,603/year

8.6.1.2 Labor Savings

In addition, eliminating the intermediate material handling would free employees to perform other tasks, resulting in labor savings. Savings from freeing one operator, at $24 per hour, would be about

24 hour/day/2 hour/spool change × 10 min/spool change × 1 hour/60 min
$$= 2 \text{ hour/day}$$

$$7 \text{ day/week} \times 50 \text{ week/year} = 350 \text{ day/year}$$

$$2 \text{ hour/day} \times 350 \text{ day/year} \times \$24 \text{ /hour} = \$16,800\text{/year}$$

8.6.1.3 Reduction of WIP Inventory

Based on our observations, we estimate that a buffer of about ten spools of drawn wire is kept to feed the "Marathon Pack" machines. At 1,500 pounds of product per spool, the value of this inventory is about

$$10 \text{ spools} \times 1,500 \text{ lb/spool} \times \$0.66\text{/lb} = \$9,900$$

If, instead of keeping inventory on hand, the value of this inventory were invested with a 10-percent rate of return, the lost income would be about:

$$\$9,900 \times 10\% = \$990$$

8.6.1.4 Other Savings

While intermediate spools are being changed, the machine is shut down. Upon start-up, scrap is created until the drawing machine hubs achieve operating temperature. Eliminating the intermediate spooling operation would eliminate this source of waste.

   Multiple handling of in-process material increases the opportunity for errors. By routing wire directly from the drawing to packing machine, this intermediate labeling step is eliminated, reducing the likelihood of mislabeled final products.

8.6.1.5 Total Cost Savings

Total cost savings from eliminating the intermediate spooling operation on one drawing/patterning cell would be about:

$$\$21,603\text{/year} + \$16,800\text{/year} + \$990\text{/year} = \$39,393\text{/year}$$

8.6.2 Estimated Implementation Cost

Management estimates that a temporary setup would cost about $2,000 per line and final equipment would cost about $10,000.

8.6.3 Simple Payback

The simple payback would be about

$$\$10,000/\$39,393/\text{year} \times 12 \text{ months/year} = 3 \text{ months}$$

| | Annual Savings | | | | |
	Resource	CO$_2$ (lb)	Dollars	Project Cost	Simple Payback
Product output	545,518 lb		$21,603		
Labor	700 hours		$16,800		
Inventory			$990		
Net			$39,393	$10,000	3 months

Reducing inventory is an important goal of the Lean organization. Carrying inventory has many costs associated with it. Obvious costs include capital tied up in inventory and the associated loss of interest on that capital, loss due to material handling damage, increased labor costs for material handling, and increased space and storage requirement. A cost from excess inventory that is not so obvious is quality. In fact, many companies have seen quality improvements resulting from inventory reductions while not focusing on quality. The reasoning is that if an upstream process is producing parts on a machine and defects occur halfway through the batch, in an organization with low levels of inventory the next downstream process will discover the defects sooner. An organization with low inventory levels can stop the process when the defect is discovered, throw out the defective inventory, and request the previous process to start another batch. The organization with lower inventory levels will also be more effective at determining what caused the defect because the batch that the defect occurred in is fresh in the minds of both production and maintenance.

9

Manage Production and Inventory Costs

9.1 Theory of Constraints: "The Drum Is a Bottleneck"

Constraints are limitations on action. They set boundaries on solutions. Yet, those boundaries have the potential to inspire business solutions. Many creativity techniques purport to free people from the constraints they place upon themselves, and in doing so, they set new boundaries within which to reflect. For example, in one creativity technique, participants are asked to think through an analogy or metaphor. Rather than thinking about potential resistance to change for implementing Lean manufacturing, they think about children's resistance to vegetables. In another, they reverse the problem. Rather than trying to come up with ways to reduce defects, they list ways to increase them. In both techniques, the idea is to open up people's minds to new ways of thinking about the problem by posing another problem with its own constraints and limitations. Altering our understanding of constraints removes the terror of a blank page. It offers us opportunities for learning about them and reinterpreting their meaning. It provides a starting point for exploring possibilities. However, starting from a different place is not the only way to understand constraints.

Another practical application of operational efficiency is the *theory of constraints* (TOC) or *constraints management*. Developed in the 1980s by Eli Goldratt, it has been applied to many aspects of Lean manufacturing, including production and supply chain management. Based on the notion that a chain is only as strong as its weakest link, TOC questions the basic notion that an increase in efficiency in any department or process decreases a firm's costs. Suppose, for example, that an increase in work-in-process efficiency leads to a buildup of finished inventory without a corresponding increase in shipments. Quite conceivably, the improvement in efficiency actually increases rather than decreases the firm's costs by causing bottlenecks in the system.

For a bottleneck work center, capacity is less than demand placed on resource. Generally, a moving bottleneck is caused by batch sizes that are too large. What happens is that a large batch scheduled on a machine or resource, which, on the average has excess capacity, prevents other products from being completed that also need the same resource. This interrupts the flow and starves downstream resources. From their perspective looking upstream, they see that particular resource as the bottleneck. However, days or weeks later, because of the product mix, this apparent bottleneck will disappear. Another large batch size somewhere else in the system will appear that does the same thing, that is, starves downstream operations.

TOC is a management philosophy with the goal of making money. According to this philosophy, every profit-making organization must have at least one constraint, which prevents the system from achieving a higher performance. These constraints must be identified and treated carefully. TOC seeks to manage production through the *drum–buffer–rope* approach. In essence, the chain is viewed as a whole rather than a collection of links. Rather than to maximize the output of every department, the goal is to coordinate the entire process to coincide with the constraints or bottlenecks. The objective is to keep the bottleneck operating at 100 percent and for nonbottlenecks to be linked to the drum. The "focusing step" of Goldratt's theory of constraints is to identify system constraints.

For a nonbottleneck work center, capacity is greater than demand placed on a resource. A nonbottleneck can become a bottleneck when it is scheduled with a batch size that is too large. For example, assume that machine 1 provides work to machine 2 and machine 3. Say that machine 1 works seven hours out of each eight hours and so is not a bottleneck. Suppose, however, that someone decides to save some setup time by scheduling work on machine 1 in much larger batches, say, twenty hours for machine 2 and fifteen hours for machine 3 (five times larger batch sizes). Machine 3 will be starved for work since it will be dealing with a forty-hour cycle rather than an eight-hour cycle, and will have to wait until machine 1 produces the parts that it needs. Thus, from machine 3's point of view, machine 1 has become a bottleneck.

In TOC, the drum is a bottleneck. It is referred to as the drum because it strikes the beat that the rest of the system uses to function. The buffer is inventory that is provided (typically prior to the drum) to make sure that the drum always has something to do. Buffers are also used to make sure that throughput is maintained throughout the production system. The rope is upstream communication from the bottleneck so that prior workstations only produce the materials needed by the drum. This keeps work-in-process inventories from building up.

9.2 Identifying and Exploring Constraints

Many design disciplines recognize and accept constraints as fundamental to their processes. While system engineers may work very hard to circumvent potential contradictions between performance and reliability, they recognize that a key component of their skill is their ability to provide creative solutions for their customers' programs taking many limitations into consideration. Engineers accept the physical limits of nature just as writers work within the tradition of their chosen literary genres and poets write haiku or sonnets. Adherence to constraints requires designers to be more creative rather than less, often enabling brilliance or beauty to emerge.

In *The Design of Everyday Things*, Norman (1990) praises constraints and demonstrates how they make it easier for people to use unfamiliar objects. If the bolt only fits into one nut, that's probably the nut it belongs in. Narrowing enables effective design by reducing the potential for error and clarifying the possibilities for action.

In *Zen and the Art of Motorcycle Maintenance*, Pirsig (1974) describes how Phaedrus helped students think of something to write. Rather than open up their options, he closed them down. He had one student write about the upper-left-hand brick on the front of the Opera House in Bozeman, Montana. He had others write about their thumbs and one side of a coin. Narrowing enabled expansiveness by providing a starting point and a focus for creating.

TOC is an overall management philosophy that aims to *continually achieve* more of the goal of a system. If that system is a for-profit business, then the goal is to make more money, both now and in future. TOC consists of two primary collections of work:

1. The Five Focusing Steps and their application to operations

2. The Thinking Processes and their application to project management and human behavior

According to TOC, every organization has one key constraint that limits the systems performance relative to its goal. These constraints can be broadly classified as either an internal constraint or a market constraint. To manage the performance of the system (the firm), the constraint must be identified and managed correctly according to the Five Focusing Steps.

9.3 The Five Focusing Steps and the Continuous Improvement Goal

TOC is based on the premise that the rate of revenue generation is limited by at least one constraining process (i.e., a bottleneck). Only by increasing throughput (flow) at the bottleneck process can overall throughput be increased.

The key steps in implementing an effective TOC approach are (see also Table 9.1):

1. Articulate the goal of the organization. Frequently, this is something like "Make money now and in the future."

2. Identify the constraint (the thing that prevents the organization from obtaining more of the goal).

3. Decide how to exploit the constraint (make sure the constraint is doing things that the constraint uniquely does, and not doing things that it should not do).

TABLE 9.1

The Five Steps of Focusing

Five Focusing Steps	Steps Expressed in Terms of Continuous Improvement	How to Implement
1. Identify the system's constraints.	What to change?	Use the effect–cause–effect method to identify constraints.
2. Decide how to exploit the system's constraints.	What to change to? Construct simple practical solutions.	Use the Evaporating Cloud method to invent simple solutions.
3. Subordinate everything else to the above decision.	How to change? How to overcome the emotional resistance to change?	Use the Socratic method to induce people to invent solutions. The Socratic approach reduces or eliminates the emotional resistance to change and allows the inventor to take ownership of the idea.
4. Elevate the system's constraints.		
5. If a constraint has been broken, go back to step 1, but do not allow inertia to cause a system's constraint.		

4. Subordinate all other processes to the decision made in step 3 (align all other processes to the decision made in step 3).

5. Elevate the constraint (if required, permanently increase capacity of the constraint; "buy more").

If, as a result of these five steps, the constraint has moved, return to step 1. Don't let inertia be the constraint ; this is the process of ongoing improvement.

As discussed in Section 9.2, the drum–buffer–rope conceptualization is often used to deal with bottleneck operations. The "drum" is the schedule; it sets the pace of production. The "buffer" refers to potentially constraining resources outside of the bottleneck, and the "rope" represents synchronizing the sequence of operations to ensure effective use of bottleneck operations.

9.4   The General Application of Theory of Constraints to Operations: Drum–Buffer–Rope Approach

TOC is most effective in reducing costs in the production, marketing, and distribution phases. Within manufacturing operations and operations management, the solution is to pull materials through the system, rather than push them into the system.

- Drum–buffer–rope (DBR)
- Simplified drum–buffer–rope (S-DBR)

Note: There are a number of different interpretations of the drum, the buffer, and the rope as some consultants and academics in this area have tried to differentiate themselves.

Drum–buffer–rope is a manufacturing execution methodology, named for its three components:

1. The drum is the physical constraint of the plant: the work center or machine or operation that limits the ability of the entire system to produce more. The rest of the plant follows the beat of the drum. They make sure the drum has work and that anything the drum has processed does not get wasted. Figure 9.1 illustrates the concept of the drum. According to the illustration, if all three links in the overall production process work to full capacity, the process will be inefficient because each link in the chain has a different capacity. Because production has the lowest capacity, the solution is to work production (the drum) at full capacity and to link raw materials and shipments to the drum.

2. The buffer protects the drum, so that it always has work flowing to it. Buffers in DBR have time as their unit of measure, rather than quantity of material. This makes the priority system operate strictly based on the time an order is expected to be at the buffered operation. Traditional DBR usually calls for buffers at several points in the system: the constraint, synchronization points, and at shipping. S-DBR requires only a single buffer at shipping. Here, the buffer is

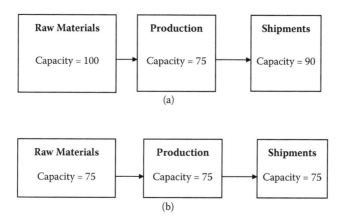

(a)

(b)

FIGURE 9.1
The concept of drum.

an allotment of time to allow for normal disruptions in operations by releasing materials before they are to arrive at the drum. Suppose the suppliers of raw materials cannot produce exactly eighty units. By placing a buffer of raw materials before the bottleneck, the throughput of eighty units can be assured.

3. The rope is the work release mechanism for the plant. Only a "buffer time" before an order is due does it get released into the plant. Pulling work into the system earlier than a buffer time guarantees high work-in-process and slows the entire system. Here, the rope refers to communications between the drum and the other operations to assure the materials are not released too soon. Clearly, it is important for all operations to be coordinated with the drum. If the buffer becomes too large, costly overhead charges may be incurred. The rope as communications is analogous to production tugging on a rope so that raw materials knows it is time to release more materials.

The DBR system balances the flow of production through a binding constraint, which reduces the amount of inventory at the constraint and improves overall productivity. TOC emphasizes the improvement of throughput by removing or reducing the binding constraints, which are bottlenecks in the production process that slow the rate of output. These are often identified as processes wherein relatively large amounts of inventory are accumulating, or where there appear to be large lead times. Using TOC, the management accountant speeds the flow of product through the binding constraint, and chooses the mix of product to maximize the profitability of the product flow through the binding constraint. A nonbinding constraint is the opposite of a binding constraint, that is, a process that does not result in a relatively large accumulation of inventory or where there are no large lead times.

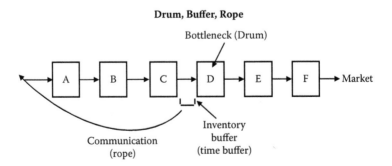

Drum, Buffer, Rope

Bottleneck (Drum)

A → B → C → D → E → F → Market

Communication
(rope)

Inventory
buffer
(time buffer)

FIGURE 9.2
Drum–buffer–rope system: a general description.

9.5 Plant Types for Operations

There are four primary types of plants in the TOC lexicon. Draw the flow of material from the bottom of a page to the top, and you get the four types. They specify the general flow of materials through a system, and they provide some hint about where to look for typical problems. The four types can be combined in many ways in larger facilities.

- I-plant—Material flows in a sequence, such as in an assembly line. The primary work is done in a straight sequence of events (one to one). The constraint is the slowest operation.
- A-plant—The general flow of material is many to one, such as in a plant where many subassemblies converge for a final assembly. The primary problem in A-plants is in synchronizing the converging lines so that each supplies the final assembly point at the right time.
- V-plant—The general flow of material is one to many, such as a plant that takes one raw material and can make many final products. Classic examples are meat rendering plants or a steel manufacturer. The primary problem in V-plants is "robbing" where one operation (A) immediately after a diverging point "steals" materials meant for the other operation (B). Once the material has been processed by A, it cannot come back and be run through B without significant rework.
- T-plant—The general flow is that of an I-plant (or has multiple lines), which then splits into many assemblies (many to many). Most manufactured parts are used in multiple assemblies and nearly all assemblies use multiple parts. Customized devices, such as computers, are good examples. T-plants suffer from both synchronization problems of A-plants (all parts are not available for an assembly) and the robbing problems of V-plants (one assembly steals parts that could have been used in another).

What is the relationship between the plant type and the theory of constraints? Here are the six necessary and sufficient questions relating to technology and the thinking process. This comes out of the haystack syndrome idea in the TOC:

1. What is the real power of the technology?
2. What limitation does it diminish?
3. What old rules helped accommodate the limitation?
4. What are the new rules that should be used now?

5. In light of the change in rules, what changes are required to the technology?

6. How to cause the change (the new win–win business model)?

TOC is appropriate for many types of manufacturing, service, and not-for-profit firms. It is most useful where the product or service is prepared or provided in a sequence of interrelated activities as can be described in a network diagram such as shown in Exhibit 5-6. The most common users of TOC to date have been manufacturing firms who use it to identify machines or steps in the production process, which are bottlenecks in the flow of product and profitability.

9.6 The Thinking Processes for Implementing Theory of Constraints

The Thinking Processes are a set of tools to help managers walk through the steps of initiating and implementing a project. When used in a logical flow, the Thinking Processes help walk through a buy-in process:

1. Gain agreement on the problem, not the solution yet.
2. Gain agreement on the direction for a solution.
3. Gain agreement that the solution solves the problem.
4. Agree to overcome any potential negative ramifications.
5. Agree to overcome any obstacles to implementation.

TOC practitioners sometimes refer to these as "the negative"→ as working through *layers of resistance* to a change. The thinking processes are outlined as follows:

- Current Reality Tree (CRT; similar to the current state map used by many organizations)—Evaluates the network of cause–effect relations between the undesirable effects (UDEs, also known as gap elements) and helps to pinpoint the root cause(s) of most of the undesirable effects.

- Evaporating Cloud (conflict resolution diagram or CRD)—Solves conflicts that usually perpetuate the causes for an undesirable situation.

- Core Conflict Cloud (CCC)—A combination of conflict clouds based on several UDEs. Looking for deeper conflicts that create the undesirable effects.

- Future Reality Tree (FRT; similar to a future state map)—Once some actions (injections) are chosen (not necessarily detailed) to solve the root cause(s) uncovered in the CRT and to resolve the conflict in the CRD, the FRT shows the future states of the system and helps to identify possible negative outcomes of the changes (negative branches) and to prune them before implementing the changes.

- Negative Branch Reservations (NBR)—Identify potential negative ramifications of any action (such as an injection, or a half-baked idea). The goal of the NBR is to understand the causal path between the action and negative ramifications so that the negative effect can be "trimmed."

- Positive Reinforcement Loop (PRL)—Desired effect (DE) presented in FRT amplifies the intermediate objective (IO) that is earlier (lower) in the tree. When the intermediate objective is strengthened it positively affects this DE. Finding out PRLs makes FRT more sustaining.

- Prerequisite Tree (PRT)—States all of the intermediate objectives necessary to carry out an action chosen and the obstacles that will be overcome in the process.

- Transition Tree (TT)—Describes in detail the action that will lead to the fulfillment of a plan to implement changes (outlined on a PRT or not).

- Strategy & Tactics (S&T)—The overall project plan and metrics that will lead to a successful implementation and the ongoing loop through POOGI (process of ongoing improvement).

9.7 Learning as Actively Addressing Constraints

A focus on constraints does not imply acquiescence to them. Rather, it implies sufficient identification and understanding to make choices. What are the constraints? Should they be accepted? Should they be negotiated? Should they be resisted? Should they be ignored? Do they matter? Answers to these questions are only made possible with adequate understanding. Understanding is an outgrowth of learning, which emerges from the interaction of a stimulus and the mind of the learner. Hence, an ability to learn is fundamental to the capacity to focus on and explore the role of constraints.

For most cognitive theorists, individual learning falls along a continuum—from reproductive thought to productive thought, from assimilation to accommodation, from accretion to structuring, from exploitation to exploration. For example, there are two strategies for problem solving: reproductive thought and productive thought. Reproductive thought occurs when people encounter familiar situations. They tend to apply previously successful strategies to deal with them. When they tackle a situation that does not correspond to the rules in their repertoires, they engage in productive thought and develop new rules to deal with it. Productive thought generates learning, which occurs when problems, or most of their elements, are new. Learning happens when people are forced to restructure and reshape previous experiences in order to deal with the current situation.

Individual learning is a process of adaptation. Adaptation takes place when an individual is changed as a result of interacting with his or her environment so that further interactions benefit the individual. Adaptation consists of two interrelated components: assimilation, the process by which the elements of the environment are changed in accord with, and are incorporated into, the organism; and accommodation, when the organism must adjust itself to environmental elements. Assimilation and accommodation interact with each other. The analogy of food consumption can be used to describe this. The mouth changes the shape of the food by chewing it in order to assimilate it into the body, but at the same time, the mouth itself changes shape to accommodate the food. In the case of information processing, assimilation incorporates information into existing cognitive structures, while accommodation rearranges, redefines, or develops understanding to interpret and incorporate new or contradictory information. There are five steps in TOC analysis:

1. Identify the constraint. Use a flow diagram. The constraint is a resource that limits production to less than market demand.

2. Determine the most efficient utilization of each constraint. Product mix decision: based on capacity available at the constraint, find the most profitable product mix. Maximize flow through the constraint: reduce setups, reduce lot sizes, and focus on throughput rather than efficiency.

3. Maximize the flow through the constraint. Drum–buffer–rope concept: maintain a small amount of work-in-process (buffer) and insert materials only when needed (drum) by the constraint, given lead times (rope). All resources are coordinated to keep the constraint busy without a buildup of work.

4. Increase capacity on the constrained resource. Invest in additional capacity if it will increase throughput greater than the cost of the investment. Do not move to investment until steps 2 and 3 are complete, that is, maximize the productivity of the process through the constraint with existing capacity.

5. Redesign the manufacturing process for flexibility and fast throughput. Consider a redesign of the product of production process, to achieve faster throughput.

One could argue that any step could be the most important; for example, step 1 can be considered to be the most important because the analysis undertaken is intended to improve the speed of product flow through the constraint.

9.8 Learning to Negotiate the Meaning of Constraints

According to community of practice theory (Wenger, 1998), learning provides the first and foremost ability to negotiate new meanings. An individual can negotiate personal meaning along a continuum from reproductive to productive thought or from accommodation to assimilation. The negotiation of the meaning of constraints fits into the same continuum. Constraints can be challenged or accepted. Even the acknowledgment of what constitutes a constraint is negotiated by the individual.

For example, someone from Los Angeles or New York has a vastly different understanding of how much traffic constitutes a constraint than does someone from Boise, Idaho, or Marion, Iowa. The negotiation of the meaning of constraints is antecedent to the choices afforded along this same continuum. Constraints can be accepted or challenged, adopted or explored.

How organizations deal with regulatory constraints provides an illustration of the meaning negotiation that is possible. When an organization faces a regulatory constraint, it has several options. It can attempt to obtain a variance for its particular situation, it can accept the constraint as it stands, it can lobby to have the constraint changed, or it can work around it. Organizations regularly get tax abatements to locate in particular states or municipalities. The law defining how many containers of what dimensions a long haul truck may pull has changed several times over the last quarter century, solely because of lobbying by organizations that haul freight or have freight to haul. Urban developers have worked around building height restrictions by creating a market for air rights.

TOC emphasizes the improvement of throughput by removing or reducing the constraints, which are bottlenecks in the production process that slow the rate of output. These are often identified as processes wherein relatively large amounts of inventory are accumulating or where there appear to be large lead times. Using TOC the management accountant speeds the flow of product through the constraint, and chooses the mix of product so as to maximize the profitability of the product flow through the constraint.

While learning is first and foremost an individual activity, the organizational milieu affects the nature and degree of learning it affords. For the most part, constraints are not the purview of individuals. Their meaning and import is explored by groups of people—the sponsor, the users, and the IT department; the client, the stakeholders, and the consultants; or the government, the engineers, and the company's managers. If the organizational environment encourages or demands the exploration of constraints, and provides the time and impetus to do so, it may be that constraints become explicit and well understood, rather than unspoken and possibly overlooked.

Unlike those in the design disciplines, managers rarely explore constraints. Instead, they expend energy to work around or eliminate them. They talk endlessly about thinking outside the box, rather than taking the time to confront the box. By investigating how other disciplines understand and work with constraints, we may be able to provide insight into how managers might become more attentive to them and learn how to work with them. For example, architects expend a great deal of effort developing the program for a building, making themselves familiar with the building code of the municipality within which they build. Engineers are constantly mindful of the constraints of the materials they use and the laws of nature. Just as they are modeled in linear or integer programming, constraints form a space from which the design must emerge.

Goldratt (1990) believes that in every organization there is a single constraint that inhibits the organization from meeting its goals. He exhorts organizations to find that constraint and then determine how to release it. He also encourages organizations to focus only on the constraint, and not the resources that are not constrained. It is very hard to determine the limiting constraint and which constraint deserves the most attention.

In management, the vast majority of constraints are socially constructed, but that doesn't render them any less real or immutable. A constraint can be the willingness with which an individual appropriates new or modified tasks. It can be the date by which a project has to be completed. It can be the government regulation that prohibits an organization from selling an insurance product. While these are probably more negotiable than the law of gravity, it is an understanding of the degree to which their meanings are negotiable and what options there are for dealing with them that determines the constraint space.

I propose that managers can learn from those who confront constraints in other disciplines. They can learn how engineers, architects, physicists, and artists identify and negotiate the meanings of constraints and how they decide to challenge them, accept them, or leverage them in the design process. By learning from other disciplines, we may develop new mechanisms to help managers think of their work as designing around and through the constraints they inevitably face.

9.9 Accounting Systems for Lean Manufacturing

TOC is most useful where the product or service is prepared or provided in a sequence of interrelated activities as can be described in a network diagram. The most common users of TOC to date have been manufacturing firms who use it to identify machines or steps in the production process that are bottlenecks in the flow of product and profitability. TOC helps businesses to determine

- What to change—What is the leverage point?
- What to change to—What are the simple practical solutions?
- How to cause the change—How do you overcome the inherent resistance to change?

Under the cost accounting view of manufacturing, all machine and labor resources should be utilized to the maximum. By contrast, the TOC concentrates key bottlenecks using DBR techniques, even if that means underutilizing nonbottleneck areas.

The drum–buffer–rope (DBR):

Drum—The bottleneck determines the speed at which the system works.

Buffer—Both materials and time act as buffers.

Rope—The buffer should not become too big or too full, because then there would be a lot of work in process. The bottleneck needs to limit the size of the buffer. Therefore, the bottleneck signals to its suppliers when it has consumed an item and therefore should receive one item.

Goldratt defines the effect–cause–effect method as the process of speculating a cause for a given effect and then predicting other effects from the same cause. The evaporating clouds method involves examining the foundations of the system. The idea is to find the minimum number of changes that are needed to create an environment where the core problem cannot exist. You do not try to solve the problem, but instead cause the problem not to exist.

9.9.1 Minimal Path Sets

Any system can be represented as a parallel arrangement of series structures. To show this, some definitions are needed:

Path vector—State vector x such that $\varphi(x) = 1$

Minimal path vector—Path vector x such that $\varphi(y) = 0$ for all $y < x$

Minimal path set—If x is a minimal path vector, then the set $A = \{i : x_i = 1\}$ is a minimal path set (minimal set of functioning components that will ensure functioning of the system).

Let $A_1, \ldots, A_s$ denote the minimal path sets of a system. Define $\alpha_j(x)$, the indicator function of the jth minimal path set, by

$$\alpha_j(x) = 1 \text{ if all components of } A_j \text{ are functioning; 0 otherwise}$$

$$= \prod_{i \in Aj} x_i$$

A system will function if all of the components of at least one minimal path set are functioning. Hence,

$$\varphi(x) = \max_j \alpha_j(x) = \max_j \prod_{i \in Aj} x_i$$

The minimal path sets define the "success modes" by which the success will be assured.

TABLE 9.2

Global Performance Measurements and Operational Performance Measurements

Global Performance Measurement	Operational Measurement
Net profit—Measurement in dollars.	Throughput—The actual rate of sales generated by the system.
Return on investment—Generally a percent of the investment.	Inventory—All the money invested in things that are intended to be sold. This includes raw materials, equipment, and so forth, but at the cost price, less any depreciation (which is operating expense).
Cash flow—The amount of cash available for day-to-day operations. From an accounting standpoint, deductions such as for depreciation are added back in since depreciation is not really money spent.	Operating expense—Money spent to convert inventory into throughput. This includes direct and indirect labor, materials, depreciation, and administrative costs.

9.9.2 Socratic Method

The Socratic method is discussed as a way to overcome resistance to change. Using this approach involves asking questions that help people invent their own solutions. According to Goldratt, when you provide people with the answers, you block them from the opportunity of inventing the answers for themselves and create emotional resistance to acceptance and implementation.

As shown in Table 9.2, there are three goals for manufacturing according to TOC:

1. Throughput, the rate at which the system generates money through sales
2. Inventory, all the money invested in purchasing things the system intends to sell
3. Operating expense, all the money the system spends in turning inventory into throughput

Traditional accounting methods work with such things as standard costs, allocation of burdens (which may consist of indirect labor, administrative costs, insurance, taxes, depreciation, etc.), and gross profits, net profits, cost centers, and profit centers, all of which may be based on standards and allocations. These may not have any basis in reality.

Many examples can be cited where traditional accounting forces wrong decisions. For example, a vice president of manufacturing is terminating a very profitable product line because, on paper, the product line is losing money. Why? Because overhead is allocated based on the amount of direct labor and this product line is almost all direct labor, with very little equipment involved. His allocation of everyone else's burden creates a paper loss.

Under these circumstances he would lose his annual bonus and be rated down in his performance. Other areas of accounting differences: inventory in traditional accounting is carried the same as cost of goods sold, that is, with all the labor and burdens included. In operational measurement's terminology, inventory is carried as the cost of the raw materials.

The primary complaints against accounting departments have to do with the fact that accounting systems measure the wrong things, are inflexible, and reward counterproductive or dysfunctional behavior. Accounting systems conform to rigid guidelines established by the Generally Accepted Accounting Principles (GAAP). As such, accounting data are often not useful for accomplishing the superordinate goals of the firm. An example is machine utilization. Machine utilization measures the proportion of time that a machine is in use. In an accounting sense, high machine utilization is preferable because it means that the investment in the machine is producing a return. From an operations point of view, this behavior results in high work-in-process inventory. Another example is quality. The generally accepted accounting definition of quality is that of conformance. However, manufacturing may desire to adopt a definition of quality that considers customer needs. Accounting would be unable to accept the latter definition as it is more difficult to quantify. The two alternative definitions of quality will reward different behavior within the firm.

Only by increasing throughput at the bottleneck process can overall throughput be increased, suggests Goldratt and DBR is the methodology used to manage maximum throughput. At the same time the organizations should reduce inventory and operating expense.

Manufactured goods can be inventoried; services cannot. There are similarities that make some management techniques usable in either environment, like the just-in-time (JIT) technique. TOC is a management philosophy that has been successfully used in manufacturing. TOC and JIT concepts emphasize continuous improvement by systematically removing waste from the system. However, from a practice perspective these concepts are quite different. JIT's practice elements are applicable in repetitive production or assembly operations. TOC emphasizes a scheduling algorithm that balances the flow of work with bottlenecks. While JIT places emphasis on zero inventories, TOC allows for inventory buffers at bottleneck operations.

Overall, the expected costs of implementing these changes include training, equipment, and increased wages due to cross-training. This is overshadowed by the expected benefits of implementing the recommendations. More throughput due to increased capacity leads to the ability to handle more customer orders that will increase revenue. In addition, decreasing both inventory costs and work in process will have a positive effect on the bottom line. An additional positive outcome of this new system is the increased customer satisfaction that will lead to customer retention and therefore increased sales.

9.10   Case Study: Marion Gyrocopter Corporation (MGC)

9.10.1   What Is the Problem?

The order volume at MGC exceeds capacity with the current system for their desired lead time. This is a problem because the company cannot meet desired lead times, which reduces customer satisfaction and loss of customer retention. This problem makes it necessary to expedite a lot of orders, which stops work in process and drives up costs. Currently, trimming has become the bottleneck of production, meaning that it is the slowest part of the production and causes work to be piled up in front of it. By increasing the capacity at trimming, throughput can be increased and Sawbones can meet order demand and deadlines.

9.10.2   How Do You Know?

There is much evidence of the problem from just taking a walk around the plant: First, there is two weeks' worth of work in process in front of trimming and one week worth of work in process in front of assembly, which is hurting the plant by tying up resources and cash, as well as hindering orders from getting delivered by the promised two-week lead time. Additionally, molding is creating extra parts to "stay efficient" with something to do, and to help build an inventory. Last, some orders are being shorted to fulfill other expedited orders, which ruin the flow when new pieces need to be molded from the beginning to make up for the shorted orders.

9.10.3   What Is Our Recommendation?

Our goal with this project is to help MGC increase throughput with a new system to accommodate desired lead times. To achieve this goal, we suggest five issues to change: adjustment of lead times, creation of a pull system, subordination of the constraint, transformation of the swing shift into a resource buffer, and job rotation.

9.10.4   How to Change?

To reach our goal, we realize it is impossible to change practices overnight. Therefore, our recommendations can be implemented in phases. The first part entails that Sawbones should increase the price of short lead-time orders to deter customers from ordering them. This will help MGC catch up on backlogged orders. Sawbones can then create a segmented market by giving customers with longer lead times a lower price on their order and charging more for shorter lead times. For example, customers who have the ability to plan ahead and can allow for a longer lead time should be given a lower rate, and orders that require a shorter lead time will cost more.

In conjunction with this, we suggest adjusting prices to when customers need it, rather than when MGC can produce it. Shorter need times will equal a higher price, and longer need times will be cheaper. Orders that have longer need times can be processed early if work needs to be fed to the buffer, which means that some orders can be delivered earlier than promised.

This relates to our next suggestion of creating a pull system, which means Sawbones should focus on production for orders and not on building up inventory. This can be implemented by producing only for customer orders with a small safety stock and not letting molding work when there are no orders to be produced. Additionally, orders need to be kept together during production, which will help minimize the orders that need retouching due to shortages.

The next phase should be implementation of a DBR system through subordination of Sawbones's constraint, which currently lies at trimming. Since trimming is the bottleneck, it is necessary to plan all operations around it and keep trimming running at full capacity. Right off, Sawbones needs to have a full-swing shift work on trimming to reduce the work in process and get rid of the backlog. After Sawbones gets on track with all of its orders, we suggest that it keep a buffer of work in front of trimming to ensure that trimming always has work to do. In our process model, we suggest keeping eight hours' worth of work in process in front of trimming, and operating molding in relation to the speed of trimming.

Additionally, MGC can turn the swing shift into a resource buffer, so that when orders throw off the flow of production, the availability of a swing shift can help return production to normal. Swing shift will just depend on what things need to be done that cannot be done during normal production hours.

As mentioned before, swing shift should focus on trimming to deplete the work in progress. To implement this, swing molders and assembly workers will need to be trained for trimming. We recommend that Sawbones encourage and provide training opportunities for employees to become cross-functional in different departments, which will help create flexibility in the workplace to adjust to variation in order sizes and types.

9.10.5 What Is Your Expected Result?

First, the most visible change will be cycle time for production. In the process model, we used two production hours as a unit to compare the as-is and to-be systems. Figure 9.3 shows the improvement from our recommendations, as implemented in the process model.

Already, MGC can expect a faster cycle time with our to-be model. The time it takes for two hours of production work took 16.9 days in the past to

Total time it takes to have two production through the system
Current system in days (8 hours)
16.9
Proposed system in days (8 hours)
3.2

FIGURE 9.3
Cycle time reduction from 16.9 days to 3.2 days.

get through the system, and is improved to just 3.2 days to get through with our new system. These results show that the new system can handle a fast throughput, which in turn, means a higher capacity.

Additionally, segmenting the market should initially help stop the onslaught of short lead-time orders and help MGC catch up on late orders. After the backlog is gone, customers who do not require short lead times have the ability to plan ahead, which leads to a lower price. Segmenting the market will help reduce work in process, increase customer satisfaction from on-time deliveries, increase revenue from short lead-time orders, and allow for greater flexibility toward planning around the bottleneck. The costs will come from dissatisfaction from customers wanting orders right away during the first phase of the change. Shorter lead times will bring in more revenue and the ability to plan ahead will help Sawbones process more orders and increase throughput. By instituting a pull system at Sawbones, we hope to lower inventory and work in process, which in turn will lower costs.

Benefits of TOC are keeping the constraint at full capacity and reducing the variability and lead time by thirteen days. This will also increase throughput. One cost of implementing a TOC system is wage differences in having to keep trimming at full capacity. Another potential cost will occur in reorganization of the plant through implementation of the new system.

The expected result of reorganization of the swing shift as a resource buffer is having more control over variability, which leads to less work in process and more throughput.

We expect more flexibility in each department due to the cross-training of employees. If a new bottleneck is created, cross-training of employees to maximize each department will help to decrease the effect. Additionally, increasing job rotation has been shown to increase productivity and job satisfaction among employees. The costs associated with job rotation would be training employees to be cross-functional as well as the possible cost of additional equipment.

9.11 Conclusion about the Case Study

Overall, the expected costs of implementing these changes are training and equipment as well as increased wages due to cross-training. This is overshadowed by the expected benefits of implementing our recommendations. More throughput due to increased capacity leads to the ability to handle more customer orders, which will increase revenue. In addition, decreasing both inventory costs and work in process will have a positive effect on the bottom line. An additional positive outcome of this new system is the increased customer satisfaction, which will lead to customer retention and therefore increased sales.

10

Kanban: Align Manufacturing Flow with Demand Pull

This chapter will explain what a Kanban system is, how it works, and how it can be implemented. The theory will then be applied to a Lean manufacturing cell to illustrate the implementation. This chapter also presents the modeling of a Kanban manufacturing and production system through the use of simulation. A model is constructed for a discrete, noncontinuous simulation of a multistage, dual-card Kanban production system. The performance of the model is monitored by tracking work in process (WIP), orders completed, and the production throughput in the just-in-time (JIT)/Kanban production environment.

Simulation helps to select the right Kanban technique to implement in a given manufacturing process. Simulation techniques also provide valuable information for anticipating production capabilities of a Kanban system before actual implementation. This can be a helpful tool for engineers, managers, and executives working with an existing Kanban system or making plans for the implementation of one.

10.1 Kanban-Based Systems

Kanban is a Japanese term meaning "signal." The term is used worldwide today to denote a form of replenishment signal used to transmit information generally regarding the movement or production of products. In general context, it refers to a signal of some kind. Thus, in the manufacturing environment, Kanbans are signals used to replenish the inventory of items used repetitively within a facility. The Kanban system is based on a customer of a part pulling the part from the supplier of that part. The customer of the part can be an actual consumer of a finished product (external) or the production personnel at the succeeding station in a manufacturing facility (internal). Likewise, the supplier could be the person at the preceding station in a manufacturing facility. The premise of a Kanban system is that material will not be produced or moved until a customer sends the signal to do so.

A Kanban system can signal the authorization to move material or product from the supplying location to the consuming location. They can also be used to signal the authorization to produce additional product. The Kanban method is the most common implementation of pull production systems. In a Kanban system, production is authorized not scheduled (in contrast with a push production method). Here, production is triggered by demand. When a part is removed from the last inventory point (or final product) of the line, the last workstation is given authorization to produce another part to replace the part that was just withdrawn. This workstation sends an authorization to the upstream workstation to produce the material that will replace the material just used to produce the new part. Each workstation repeats this process:

- Send an authorization signal (or Kanban) to produce a part to the upstream workstation
- Replenish the part just used by the downstream workstation

One benefit of a Kanban system is that it puts limits on inventory buildup. A Kanban acts as a limit. When the Kanban is full, no additional product can be made, or moved, into that location. Putting limits on inventory has some big benefits on business bottom line, including less cash is tied up, less space, less handling, and less handling damage.

Think of items in WIP as standing in line, waiting their turn for processing. The longer the line, that is, the more WIP, the longer they will have to wait. Reductions of WIP inventory have the additional benefits of reducing products' lead time. Fewer inventories also reduce the amount of scrap or rework required when a defect is discovered. And, since fewer inventories mean shorter lead times, a Kanban control system shortens the time between creation of a defect and its discovery, thereby improving the chances of correctly diagnosing its cause. For repetitive items, a Kanban system can also reduce the reliance on forecasts. One of the most powerful aspects of a Kanban is the ease that it provides for forcing continuous improvement at the grassroots shop-floor level of the organization.

Inventory reduction exposes problems and forces solutions to those problems. A Kanban system provides a simple visible mechanism for shop-floor people to translate top-level management objectives into concrete actions, for example "Cut the size of the Kanban between operation 1 and operation 2 by 70 percent in the next six weeks." In the Kanban system, a workstation requires materials and a Kanban to start the production of the new part. A schematic for a four-cards Kanban system is presented in Figure 10.1. In this figure the circles represent the workstations. In Figure 10.1, the Kanban limit is four units. Note that with this simple control mechanism, a limit has been set on inventory, space, lead time, and potential defects.

Note also that the Kanban mechanism provides a simple, powerful mechanism to force continuous improvement. Think of the Kanban quantity as a buffer that allows (hides) problems to exist. How do you force those problems to be exposed? Reduce the Kanban quantity! If everything is running

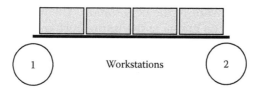

FIGURE 10.1
Kanban limits, four units.

well at a Kanban size of four, cut it to three. Some of the main operational rules for a Kanban system include:

- Downstream operations withdraw only the quantity of items they need from upstream operations. The quantity is controlled by the number of cards.
- Each operation produces items in the quantity and sequence indicated by the cards.
- A card must always be attached to a container. No withdrawal or production is permitted without a Kanban.
- Only nondefective items are sent downstream. Defective items are withheld and the process is stopped until the source of the defect is remedied.
- The production process is smoothed to achieve level production. Small demand variations are accommodated in the system by adjusting the number of cards.
- The number of cards is gradually reduced to decrease WIP and expose areas of waste.

The typical Kanban signal is an empty container designed to hold a standard quantity of material or parts. When the container is empty, the customer sends it back to the supplier. The container has attached to it instructions for refilling the container such as the part number, description, quantity, customer, supplier, and purchase or work-order number. Some other common forms of Kanban signals are supplier replaceable cards for cardboard boxes designed to hold a standard quantity, a standard container enclosed by a painting of the outline of the container on the floor, and color-coded striped golf balls sent via pneumatic tubes from station to station.

The Kanban system described is a pull system. Traditionally, a push system is and has been employed. The push system is also more commonly known as the materials requirements planning (MRP) system. This system is based on the planning department setting up a long-term production schedule, which is then dissected to give a detailed schedule for making or buying parts. This detailed schedule then pushes the production people to make a part and push it forward to the next station. The major weakness of this system is that it relies on guessing the future customer demand to develop the schedule that production is based on and guessing the time it takes to produce each part. Overestimation and underestimation may lead to excess inventory or part shortages, respectively.

One of the major reasons Kanbans are used is to eliminate or reduce the aforementioned wastes throughout an organization due to the pull system that is employed. Waste can come from overproduction (inventory) and, therefore, the need for a stockroom. This waste is eliminated. Part shortages (underproduction) are also eliminated. Costs are reduced by eliminating the need for many of the purchasing personnel and the paperwork associated with purchasing. The planning department's workload is also reduced as it no longer needs to produce work orders.

10.2   Why Use Kanban?

Kanban serves two functions for a company: to help control production and to help improve processes. To help control production, Kanban ties different manufacturing processes together, ensuring that the necessary amounts of material and parts arrive at the appropriate time and place. This guarantees that only the required amounts of parts are being used and there is no excess inventory in process. To help processes, Kanban improves operations that are being used in production by emphasizing reduction in inventory costs.

A company is able to benefit from using Kanban in numerous ways. By implementing Kanban, a company is able to reduce its inventory levels to only what is needed for each process; this helps lower inventory costs dramatically. Kanban cards also create a visual schedule for production. They state how many parts should be created and when they must be created by. This allows a company to put more emphasis on the orders that are due in the near future, such as a couple days, versus the orders that are due in a couple weeks.

The productivity of a company can increase by implementing the Kanban system. This is due to Kanban freeing people up from being overwhelmed with work. Since an operator cannot start producing parts until a Kanban card is received, the employee is free to work on other tasks or duties until he or she receives a card. If operators are doing more than just producing their required parts, more work is being done, which in turn helps increase productivity levels for the company. If productivity increases, it is only natural that lead times will also be reduced, which will help a company establish better relationships with their customers.

Kanbans serve many purposes. They act as communication devices from the point of use to the previous operation and as visual communication tools. They act as purchase orders for your suppliers and work orders for the production departments, thereby eliminating much of the paperwork that would otherwise be required. In addition, Kanbans reinforce other manufacturing objectives such as increasing responsibility of the machine operator and allowing for proactive action on quality defects. However, Kanbans should not be used when lot production or safety stock is required because the Kanban system will not account for these requirements.

10.3   How Kanban Works

A card is the most widely used way of conveying information in a Kanban system. However, there are other signals that can be used, including radio frequency identification (RFID) tags, painted or marked off areas on the ground, barcode signals, or color-coded or dedicated containers. Basically, the only true requirement is a clear, visible signal that indicates specific action.

There are two types of Kanban cards. Generally both are used in a Kanban system, but sometimes an organization may choose to only implement a one-card Kanban system. A conveyance Kanban authorizes the withdrawal and movement of a container of material. A production Kanban authorizes the manufacture of a container of material. Kanban is a system of using cards to authorize the transport and manufacture of bins of material within an organization. These cards circulate from start to finish within each workstation and then back to start again. The process periodically repeats itself, depending on how long the process takes. An example of how Kanban works can be explained by the following:

1. External demand takes a container of parts.
2. A worker downstream at workstation 2 now needs a container of parts.
3. He takes a conveyance Kanban from the container that was just emptied.
4. He finds a full container of the part needed in inventory or a WIP bin.
5. He places the conveyance Kanban in the full container and removes the production Kanban from the full container.
6. He places the production Kanban on a post at workstation 1.
7. He then takes the full container of parts with its conveyance Kanban back to workstation 2.
8. The process then cycles over and over again.

10.4   The Rules for Kanban

There are rules that must be followed for Kanban to work to its fullest potential. However, there are different approaches and variations to these rules. The six rules are:

Rule 1: Downstream processes withdraw items from upstream processes.

Rule 2: Upstream processes produce only what has been withdrawn.

Rule 3: Only 100 percent defect-free products are sent to the next process.

Rule 4: Level production must be established.

Rule 5: Kanbans always accompany the part themselves.

Rule 6: The number of Kanbans is decreased gradually over time.

Even though following all the rules is critical for the Kanban system to operate, rule 1 is considered the most important. This rule makes sure that what has been sold will be made. It makes the flow of Kanban through the workshop go smooth. Withdrawing a product basically "pulls" another product through the factory. Certain points must be remembered when following this rule. The first one is that nothing should be transported without a Kanban. Next, the operator should only withdraw as many parts as the Kanban indicates, and a Kanban must always accompany an item. Last, that a part must always follow the Kanban's indicated path.

Rule 2 makes sure that only the number of products on the Kanban card is produced. It is very crucial that this rule is followed. If an operator tries to get ahead, he or she will back up the next workstation. This is what causes bottlenecks throughout the factory.

The next rule makes sure that quality is put into every aspect of a part. An operator must make sure that the part he or she produces is 100 percent defect-free. This prevents the next workstation from having to fix what the last workstation defected. If a product does have a defect, the operator should stop the machine and try to fix the problem. He or she should never try to modify the product or continue production.

Rule 4 is leveling the load of production. This will eliminate the peaks and valleys throughout the workshop. Having a level amount of production will make sure that daily production runs smooth. This rule also allows you to adapt to small fluctuations in demand by fine-tuning production as conditions change.

Rule 5 used to be Rule 1 because of its importance. The rule is the visual control of the product in the system. These identification tags let the operator know how many products to produce and what to do with the part when it is finished. Without these tags, there would be no organization in the system.

The last rule deals with decreasing the number of Kanbans. Even though it sounds like a bad idea to decrease the amount of Kanbans, in all actuality it is not. If the number of Kanbans decreases, a lot can be learned from it.

Such problems are line-stopping problems and missing items. Decreasing the number of Kanbans is an excellent way to find out the company's optimum production level.

10.5　Steps for Implementing Kanban

The first step of implementing Kanban is forming a Kanban team. This is essential because implementing Kanban is something that cannot be done alone. At a minimum, the team should include production management/ supervision, materials management, material handlers/warehouse associates, and production operators. Once the team is formed, a leader needs to be appointed who can coordinate and lead the team. The first task of the team will be to develop a plan of attack. This plan should include a schedule, a budget, and a list of supporting employees who will have a role in implementing Kanban outside of the team. It is key to remember that the team will need the full support from management starting at the top all the way down.

The next step is collecting and analyzing data. The information that needs to be collected for each step includes number of parts produced, changeover times, downtime, and scrap levels. Take time to make sure that the data is accurate as it is important to have good data. Use the rule "good data in results in good results out, but bad data in results in bad results out." After collecting and compiling the data, check it for consistency, accuracy, and realism.

After collecting data, you can use that information to size the Kanban. This essentially means that you are calculating how many containers are needed between each station to effectively operate the Kanban and keep the customers supplied. These numbers essentially are the maximum quantity of inventory. All of the calculations basically boil down to calculating the number of Kanban containers between each station. This calculation can be found in Figure 2. The calculations required to use that calculation are found in Figure 3 and Figure 4. Once the calculations have been made, do a reality check. Do these numbers make sense? If these calculations are used, will inventory change?

The next step is to develop the Kanban. This consists of selecting the signaling mechanism, developing the rules for operation, and creating the visual management plans for the Kanban. These three steps tend to be interrelated. It is important to keep in mind that Kanban is a visually based signaling system and that the idea is that if you can visually see that a container needs to be filled, then the system will work. In addition to simply using Kanban cards there are Kanban boards. Cards are posted on the board so that anyone walking by the station can see easily what that station is working on and if they are ahead or behind schedule. It is important to remember that whatever signaling system you choose, it needs to be simple, cards should not be redundant, and the signals should be easy to manage.

Before the Kanban system implementation, it is important to spend some time training employees on why and how to use Kanban. The best way to do this is by using conference-style training sessions. If the employees do not understand how Kanban works or do not take ownership in the change to a Kanban system, success will be difficult. The following outline can be used:

I. Kanban basics

II. How Kanban will work

    A. What is the signal

    B. How will the material move

    C. Review of the rules

III. What are the scheduling decisions and rules for making the decisions

    A. Use examples of different schedule conditions to teach how and what decisions to make

IV. Discuss when to call for help and what to do specifically when encountering a red signal

V. Conduct a dry run

Now that the Kanban is calculated and designed, and the employees are trained, the Kanban system is almost ready to begin. Before starting, make sure that the Kanban signals are completed, the rules are posted, and the visual management information is posted. Finally you are ready to start. Be aware that problems will occur and the best thing is to not worry, but solve the problem and keep moving on. Four typical problems occur when implementing Kanban that you should be aware of before starting.

1. Production operators and supervisors are not sure that the Kanban started.

2. Production operators do not follow the signals.

3. No one knows what to run because there were too many cards at the same time.

4. Production operators cheat with the signals.

The first thing to keep in mind is that it needs to be clear that the Kanban is starting. This may mean posting memos, making announcements, or talking personally with each operator. The other thing is that in training, it should be ensured that everything is clear and that all questions are answered at the end of training. Reviewing with the operators before going live is a good idea.

Kanban is a Lean manufacturing principle and the underlying theme of all Lean concepts is continuous improvement. This means that upon implementation of Kanban it is important to audit the success of the changeover and adjust Kanban quantities to continually improve upon the success of the system.

10.6 Types of Kanban

10.6.1 Dual-Card Kanban

The dual-card Kanban system is more commonly referred to as the Toyota Kanban system as Toyota was the first to fully employ this system. It is a more useful Kanban technique in large-scale, high-variety manufacturing facilities. In this system, each part has its own special container designed to hold a precise quantity of that part. Two cards are used: the production Kanban, which serves the supplier workstation; and the conveyance Kanban, which serves the customer workstation. Each container cycles from the supplier workstation to its stock point to the customer workstation and its stock point, and back while one Kanban is exchanged for another. No parts are produced unless a production Kanban authorizes it. There is only one conveyance Kanban and one production Kanban for each container and each container holds a standard quantity (no more, no less).

10.6.2 Single-Card Kanban

The single-card Kanban system is a more convenient system for manufacturing facilities requiring only a few varieties in their parts. Essentially, the single-card Kanban system is simply a dual-card Kanban system with the absence of the production Kanban and designated stock points.

10.7 Kanban Development

Developing a Kanban system entails four major steps (which may be slightly modified depending on the requirements of the facility):

- Step 1 is to pick the parts to use in Kanban. In general, these parts should be used repetitively within the plant with fairly smooth production requirements from month to month.
- Step 2 is to calculate the Kanban quantity. This quantity is based on the following formula:

Kanban quantity = Weekly part usage × Lead time

× Number of locations × Smoothing factor

The weekly part usage is, as the name implies, the quantity of the part under consideration every week. The lead time is given by the supplier. The usual manufacturing facility lead time is five working days per week. The number of locations tells us how many locations should have a full container to begin with. The smoothing factor is used to account for seasonal fluctuations in demand. It is a constant determined by the ratio of the fluctuating demand to the regular demand.

- Step 3 is to pick the type of signal and container to be used to hold a standard quantity. The container should aid visual identification, ease of storage, and count of material at the point of use.

- Step 4 is to calculate the number of containers. This calculation is performed using the following formula:

$$\text{Number of containers} = \text{Kanban quantity}/$$
$$\text{Number of parts held per container}$$

This paper investigates why machines are presently designed to reduce unit labor cost by increasing the speed of the machine or by eliminating direct labor altogether with automation. Machine design practices are currently shown to be operationally focused rather than system focused. This paper illustrates the way the unit cost equation and operationally focused machine design approaches combine to result in costly factory-system implementations that do not achieve the enterprise objectives. Examples of the hidden costs that are not disclosed by the unit cost equation are then identified. As an alternative to cost management with the unit cost equation, a manufacturing system design decomposition is presented. The decomposition provides a methodology to identify each manufacturing objective and the chosen solution. The decomposition approach is used as the basis for contrasting the difference between mass and Lean production.

10.8　Kanban Design to Achieve Manufacturing System Objectives

The traditional manufacturing cost accounting system, which is now widely used as the basis for manufacturing management decisions, was developed in the 1920s by DuPont, General Motors, and others. This cost accounting approach is based on the realities of the 1920s, when direct labor was a single dominant factor of all manufacturing costs other than raw materials. Consequently, this cost accounting system typically equates "cost" with direct labor cost. All other costs are miscellaneous, then lumped together as an overhead, which is then allocated based on direct labor time.

10.8.1   Unit Cost Coupled with Operation-Focused Engineering

This traditional unit cost approach has long been the performance measure of manufacturing cost. If we combine the operation-focused engineering, which is a term that describes the design and optimization of a single manufacturing process or machine in isolation of the product flow Shingo (1989), with the unit cost equation (Equation 10.1), the departmental mass environment is the typical result.

$$\alpha = \frac{(C_{dl} + C_m + C_{oha})}{N_p} \tag{10.1}$$

where
α = Unit cost of product
C_{dl} = Direct labor cost
C_m = Material cost
C_{oha} = Overhead allocation of product
N_p = Number of parts produced

The direct labor cost, C_{dl}, can be calculated by the following equation:

$$C_{dl} = W_{dl} \times DL_p \tag{10.2}$$

where
W_{dl} = Wage of direct labor per hour
DL_p = Direct labor hours consumed by the product
The overhead allocation of product C_{oha} can be calculated by

$$C_{oha} = \beta \times C_{ohp} \tag{10.3}$$

where
β = Burden rate
C_{ohp} = Total plant overhead cost
The burden rate, β, can be calculated by

$$\beta = \frac{DL_p}{DL_{tot}} \tag{10.4}$$

where DL_{tot} is the total direct labor hours of the plant.
Capacity for each operation is calculated by

$$\mu_i = \frac{Y_i}{X} = \frac{\sum_{j=1}^{N} M_{CT_{ij}}}{X} \tag{10.5}$$

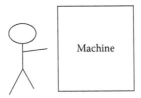

FIGURE 10.2
Traditional one machine per one operator situation.

where, for each operation *I*,
   μ = Number of machines
   Y_i = Total daily processing time
   X = Daily available operating time
   N = Number of products *j*
   $M_{CT_{ij}}$ = Machining cycle time

Each department in the plant layout corresponds to a processing operation. Furthermore, the people in this type of manufacturing system typically
operate one or at most two machines. Figure 10.2 illustrates the operation-
based processing environment in which one person operates one machine.
In this environment the unit labor cost is coupled with the production rate
of the machine.

Enterprises that use Equation 10.1 as their cost management system attempt
to reduce unit cost by determining at least three functional requirements
(FRs) that affect the mass manufacturing system design:

FR 1—Eliminate the need for direct labor: $DL_p \to 0$

FR 2—Increase the number of units/time to infinity: $N_p \to \infty$

FR 3—Reduce labor wage: $W_{dl} \to 0$

The first way is to eliminate the need for direct labor by implementing automated machines. Second, unit cost is reduced by maximizing the number of
units produced during a certain time interval through increased processing
speed of the machine. Third, unit cost can be minimized by directly reducing
the labor wage (moving plants to low-wage countries is one such approach).

10.8.2 The Manufacturing System Design Decomposition

As an alternative to the problems identified with the unit cost equation, a
manufacturing system design decomposition (Cochran et al., 1999) has been
developed based on axiomatic design (Suh, 1990). Axiomatic design is based
on two axioms, and the contention is that all good designs comply with these
two axioms. The application of these two axioms supplies the basis for objective assessment of design solutions. Axiom 1 says that the best designs are

the ones with the most independence of the functional elements. Lack of independence, or coupling, is the cause of lack of adjustability or poor control. The result is that iterations are required to satisfy the functional requirements, that is, the musts and shoulds, of the design. The second axiom is that among the independent design solutions the design with the least information content will be the best. Information in this sense is numerically equal to the log of the inverse of the probability of success, such that minimizing the information content is equivalent to maximizing the probability of success.

Applying axiom 1, the independence axiom, requires that the design have a hierarchical structure in two domains, the functional and physical. To apply axiom 1, interaction matrices between the functional and physical domains are generated at each level and for each branch in the hierarchy. Alternative design solutions are examined based on how they influence coupling between different design parameters (DPs).

In the decomposition, maximizing customer satisfaction rather than simply producing more increases sales revenue. In addition, since charging more than market price is almost impossible in today's highly competitive market, the production costs are reduced to the target cost. To achieve the target cost, all nonvalue-added costs are eliminated. Finally, to minimize production investment, right-sized machines are used instead of highly automated, high-speed machines. As shown in Figure 10.3, one of the major business FRs is FR1: Maximize return on investment. This functional requirement can be broken into the following functional requirements:

FR11: Increase sales revenue

FR12: Minimize production costs

FR13: Minimize production investment

The development of the hierarchies of functional requirements (the elements of the functional domain) and design parameters (the elements of the

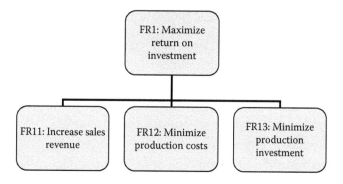

FIGURE 10.3
Decomposition of a business, functional requirement.

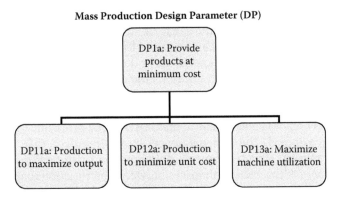

FIGURE 10.4
Decomposition of a mass production design parameter.

physical domain) requires following a sequence of activities. It is, in part, the development of the sequence of activities that creates some of the issues associated with the implementation of axiomatic design, but ultimately this sequence contributes to its benefits as well. As shown in Figure 10.4, one of the major mass production DPs is DP1a: Provide products at minimum cost. This design parameter can be broken into the following design parameters:

DP11a: Production to maximize output

DP12a: Production to minimize unit cost

DP13a: Maximize machine utilization

In axiomatic design, the first axiom stresses simplicity in design by minimizing the coupling of FRs with one other, and of unintended coupling of DPs with FRs, allowing for finer adjustment and control functions without need for iteration. Compliance with axiom 1 also avoids unintended consequences. The second axiom stresses robustness and reliability by minimizing the amount of information required for extended operation or repair. Compliance with these two axioms provides a basis for evaluating the quality of designs and can lead to important design innovations. The ability to assess design quality through axioms in this manner is the principal element that distinguishes axiomatic design from other design approaches. As shown in Figure 10.5, one of the major Kanban system DPs is DP1b: Optimize production flow. This design parameter can be broken into the following design parameters:

DP11b: Production to maximize customer satisfaction

DP12b: Produce at target cost

DP13b: Right-sized investment

With this design decomposition, the differences between equipment design in Lean and mass plants can be explained. The decomposition shows that the

Kanban System Design Parameter (DP)

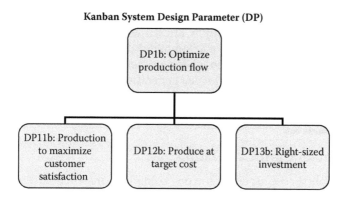

FIGURE 10.5
Decomposition of a Kanban system design parameter.

equipment in mass manufacturing systems is the result of operation-focused thinking while the equipment in Lean production systems is the result of a new system design thinking, which uncouples labor cost from the speed of the machine.

Lateral decomposition separates customer needs (CNs) from constraints (Cs) that should never be violated. Customer needs logically lead to functional requirements that should be consistent with the first axiom. Functional requirements, in turn, lead to design parameters that should be consistent with both axioms. Guidelines for separating FRs, DPs, and Cs are given in Table 10.1. Vertical decomposition separates the design into different layers of detail in a

TABLE 10.1

Guidelines for Identifying FRs, DPs, and Cs

	Functional Requirements (FRs)	Design Parameters (DPs)	Constraints (Cs)
Definition	Describes what the design should do	Describes what the design should look like	Describes to what limits the design must conform
Purpose	Satisfy customer needs	Satisfy FRs	Define boundaries
Word usage	Start with verbs (imperative phrases inspiring action)	Start with nouns (declarative phrases prescribing solutions)	
Tolerancing	Functional tolerances	Physical tolerances, e.g., $x = \pm.05$ mm	Have limits, e.g., $x < .5$ m in length
Uniqueness	Independent from other FRs	One selected to satisfy each FR	May be linked to several FRs
Relation to solution	Requires a DP	Specifies a purchased component, drawing, or a process	Must not have a DP

design hierarchy. Higher level DPs set the stage for lower level FRs. The process for developing the hierarchy is known as zigzagging between FRs and DPs, and this recursive aspect of axiomatic design is one of the elements that distinguish it most from traditional design methodologies that move sequentially from problem definition, to idea generation, to concept selection.

10.9   Kanban Card Calculation

To reduce labor cost, operator's work content is matched to the customer demand cycle time (takt time) and improvement results from decreasing the motions of the operator. The Kanban system design achieves multiple FR–DP pairs as defined by the manufacturing system design decomposition. The Kanban system design decomposition presented can help management keep its enterprise competitive by adding value to the products and enhancing customer satisfaction. A Kanban system design with objectives that do not reflect the operational focus of the traditional cost approach can serve as a guide to design machines and operate the system in a more competitive manner. Axiomatic design reveals the relationships for the functional requirements of a system and the corresponding design parameters and clearly presents them by the design decomposition procedure. The unit cost equation leads to performance measurement that drives the design of a mass production system. With the decomposition method, objectives of the enterprise drive the system design and corresponding performance measures may be derived.

Figure 10.6 displays the flow of material and production Kanbans in a process controlled by Kanban. A production Kanban is used to originate production of a predefined quantity of parts and follows the part from its initial workstation

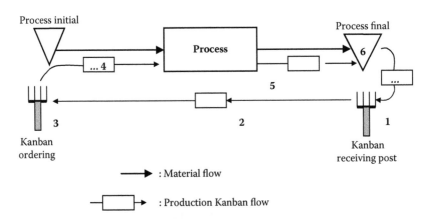

FIGURE 10.6
Flow of material and production Kanbans in a process controlled by Kanban.

to the final buffer. Note here that if the process represents a single machine then we get the classical one-machine production Kanban environment.

To determine the number of Kanbans needed for each part and for each process, it is necessary to compute the level of inventory that must be kept in the final buffer, that is, the average part quantity consumed by subsequent processes during the interval from part-production ordering to part withdrawal from the buffer. This interval is equal to the time it takes for one Kanban to complete the entire cycle.

Parts consumed during one Kanban cycle = Average demand

$$\times \text{ Kanban cycle time} \qquad (10.6)$$

Since the demand is not constant, a safety coefficient (α) is introduced to Equation 10.6.

Parts consumed during one Kanban cycle = Average demand $\times$ $(1 + \alpha)$

$$\times \text{ Kanban cycle time} \qquad (10.7)$$

As a rule of thumb, α should be kept less than 10 percent and should be decreased as the demand becomes more leveled. Since each container possesses one card, the number of Kanbans in circulation is then obtained by dividing the number of parts consumed during a Kanban cycle time by the number of parts per container:

$$\text{Number of Kanbans} = \frac{\text{Parts consumed during one Kanban cycle}}{\text{Number of parts per container}} \qquad (10.8)$$

Or, more precisely,

$$\text{Number of Kanbans} = \frac{\text{Average demand} \times (1 + \alpha) \times \textit{Kanban cycle time}}{\text{Number of parts per container}} \qquad (10.9)$$

The Kanban cycle time is the time spent by a Kanban to complete a full cycle; it involves six distinct phases as numbered in Figure 10.6. They are:

1. The production Kanban is removed from the full container when the container is withdrawn from the final buffer for transfer to a downstream process; the Kanban is then placed in the Kanban receiving post and waits there until collection.
2. The Kanban is transferred from the Kanban receiving post to the Kanban ordering post at the initial workstation.
3. The Kanban waits at the production ordering post behind the other production Kanbans (FIFO).

4. The Kanban is taken from the production ordering post and attached to an empty container, the process machines are set up, the quantity of parts to process (equal to the container size) that was indicated on the Kanban are withdrawn from the initial buffer, the parts are then processed and placed in the container upon completion.
5. The full container (with its Kanban) is transferred to the final buffer.
6. The container waits until withdrawn by downstream process (the cycle is then complete).

Equation 10.10 sums up the Kanban cycle time.

$$
\begin{aligned}
\text{Kanban cycle time} = {} & \text{Kanban waiting time in receiving post} \\
& + \text{Kanban transfer time to ordering post} \\
& + \text{Kanban waiting time in ordering post} \\
& + \text{Lot processing time (Internal setup} + \text{Run time} \\
& + \text{In-process waiting time)} \\
& + \text{Container transfer time to final buffer} \\
& + \text{Container waiting time in final buffer} \quad (10.10)
\end{aligned}
$$

Some of the waiting times, such as the Kanban waiting time in the ordering post, cannot be directly computed and are usually determined by experience. Therefore, the number of cards is initially estimated by applying the previous formulae and then adjusted according to the process's behavior. If there is too much inventory waiting in the final buffer, one or several cards may be removed. If there is a risk of stockout, the number of cards can be increased or the Kanban cycle time decreased (through process improvement).

The container size significantly affects the system's behavior. A good rule of thumb is to set the container size to 10 percent of the daily demand, thereby limiting the number of setups to ten setups per day for each part; this implies, however, that machine setup times have been reduced to single digit numbers (less than ten minutes).

EXAMPLE 10.1

Average demand = 100 parts/hour
Container size = 100 parts
Kanban waiting time in receiving post = 5 min
Kanban transfer time to ordering post = 5 min
Kanban waiting time in ordering post = 30 min
Maximum internal setup time = 5 min
Maximum processing time = 0.25 min/part
In-process waiting time = 10 min
Container transfer time to final buffer = 5 min

Container waiting time in final buffer = 10 min
Initial safety coefficient: $\alpha = 0\%$

Thus,

Kanban cycle time = 5 + 1 + 30 + [5 + (0.25 × 100) + 10] + 5 + 10 = 91 min

$$\text{Number of Kanbans} = \frac{\frac{100 \text{ parts}}{60 \text{ min}} \times 91 \text{min}}{100 \text{ parts}} = 1.52$$

Rounded up to the nearest integer, we get two Kanban cards. Based on two Kanban cards the actual Kanban cycle time is:
The difference between 91 and 120 minutes is about 25 percent of the Kanban cycle time; thus there is no need to consider an additional safety coefficient.

Kanban golden rules are summarized as follows:

- Do not move nonconforming parts to a downstream process.
- Ensure that downstream processes withdraw parts from upstream processes in the correct quantity and at the right time.
- Do not let upstream process produce more than the quantity of parts withdrawn by downstream process.
- Ensure that the production is leveled; the underlying rule to ensure leveled production is that the Kanbans should always be processed on a first-come, first-serve basis. Altering the card order will result in unnecessarily speeding up some parts and delaying other parts.
- Do not attempt to transmit large demand variations with the Kanban system.
- Balance cycle times for smooth production, and constantly improve cells and workstations.

10.10 Summary

Derived from the combination of two Japanese words, *kan* ("visual") and *ban* ("card" or "board"), *Kanban* roughly translates to "sign board" or "signal board." Kanban is a visual signal that something needs to be replenished. Lean manufacturers use Kanban to drive a process to make, move, or buy the appropriate parts. Thus, Kanban becomes one of the fundamental building blocks of a pull (or consumption based) replenishment system. No card? No replenishment.

The Kanban method helps you manage inventory or processes; it allows you to easily know what is in stock and what has been shipped because each

Kanban has a certain number of products within it. Even when you use an electronic signal you know how many of a product are shipping and how many are produced. As an example, let's assume the Kanban is an electronic signal and you are producing refrigerators. If you implement a signal for 20 refrigerators going through the installation of thermal insulation, then you know that the next production area receives 20 refrigerators.

Kanban or pull is a process that transforms a company's production strategy. A key objective of Lean manufacturing is to link production processes by continuous flow. Where desired flow is not possible due to monuments (batch dependent operations), remote operations, or unreliable processes, a manufacturer may convert production from push operations to pull—that is, from manufacturing in order to forecast to manufacturing in order to replenish units sold. Kanbans (one of several types of pull signals) are replenishment signal cards that indicate how many parts or units need to be produced to replace those that have been either consumed by downstream processes or sold to customers.

11

Jidoka: Implement Lean Manufacturing with Automation

11.1 What Is Jidoka?

Just-in-time (JIT), Toyota Production System, Lean manufacturing, Kanban, and pull system are common buzzwords that are heard on many shop floors or in war rooms throughout the manufacturing industry. One term that goes along with Japanese manufacturing practices is *Jidoka*. The Toyota Production System is often modeled as having two pillars, with one of the pillars being JIT and the other Jidoka. Jidoka has two meanings in Japanese. One of the meanings is "automation"—changing a manual process into a machine process. The other meaning is "automatic control of defects" (Monden, 1983, p. 141).

Standard and Davis (1999) translate Jidoka to mean "quality at the source," and Imai (1986) and Monden (1983) refer to it as meaning "autonomation." This meaning incorporates the insight or mind of a human to troubleshoot and correct failures. Toyota refers to this type of Jidoka as *Ninbennoaru Jidoka*, or literally translated, "automation with a human mind" (Monden, 1983, p. 141). The short definition for Jidoka would be a process or technique of detecting and correcting production defects. It always incorporates the following devices:

- A mechanism to detect abnormalities or defects
- A mechanism to stop the line or machine when abnormalities or defects occur (Monden, 1983)

The way Jidoka achieves its purposes in Toyota can be seen in Figure 11.1.

By involving the human mind in Jidoka, there are two methods to stop a line. One method involves the judgment of the line worker and the other is the machine shutting off with its automatic devices. The first method requires line workers to be empowered to shut down production to address problems. A common tool used when a line worker shuts down the line is an *andon*. An andon signals to the rest of the shop floor that something is wrong and needs to be addressed. The problem is usually addressed by a team of individuals.

A tool used for shutting machines down is *poka yoke*. Poka yoke, or fail safing, is a practice of ensuring that defects cannot occur. Examples of

---

JIDOKA – Autonomation

The principle of stopping work immediately when a problem occurs.

"Automation with a human element"

Eliminating the human factor from the system through Autonomation.

FIGURE 11.1

How autonomation attains its purposes: involving the human mind.

poka yoke can include things like automatic switches, lasers, and jigs. Whenever a defect is detected or whenever the specified amount of product is produced, a signal is sent to the machine to shut down. This practice frees up line workers to be able to manage more than one machine at a time.

An example of Jidoka in practice can be found at Toyota. The assembly line moves at a constant rate and between processes is a mat. If a worker does not finish his operations in the specified amount of time, he will step on the mat, and, in doing so, will shut down the line. This will focus attention on his area to troubleshoot why it is taking him longer than specified. Another example would be lasers and limit switches that detect the length of various parts. If the parts do not meet the requirements programmed into the system, a signal will be sent to cause the line to stop and workers will begin to address the problem.

11.2   Ladder Logic: A Tool to Implement Jidoka

Ladder logic is a graphical language for programming programmable logic controllers (PLCs), which are widely used for incorporating human elements into manufacturing automation. The application program used in PLCs is called the *ladder logic program* or the *ladder diagram.* Ladder logic input contacts and output coils allow simple logical decisions. Functions extend basic ladder logic to allow other types of control. For example, the addition of timers and counters allows event-based control. A longer list of functions is shown in basic PLC function categories. Combinatorial logic and event functions have already been covered. This chapter will discuss data handling and numerical logic. The next chapter will cover lists and program control and some of the input and output functions. Remaining functions will be discussed in later chapters.

Ladder logic was originally invented to describe logic made from relays. The name is based on the observation that programs in this language

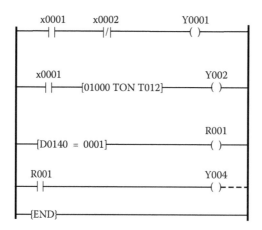

FIGURE 11.2
A simple ladder logic program.

resemble ladders, with two vertical "rails" and a series of horizontal "rungs" between them. There are two power rails in a ladder logic program. These are the two vertical lines in the program, one on the left and the other on the right, as shown in Figure 11.2. The Ladder logic diagram looks like a ladder. It's more like a flow chart than a program. Between the vertical lines are rungs of conditionals on the left that lead to outputs on the right. Each component in a ladder logic program is called an *instruction*. The instruction or instructions connected in parallel with other instruction(s) create a *branch*. Each part of the ladder logic program, which is a complete horizontal line and its branch(es) between the two power rails, is called a *rung*. The relay logic instructions include

- --( )-- A regular coil, true when its rung is true
- --(\)-- A "not" coil, false when its rung is true
- --[ ]-- A regular contact, true when its coil is true (normally false)
- --[\]-- A "not" contact, false when its coil is true (normally true)

So the first rung of Figure 11.2 means that if input 1 is energized and input 2 is not, then energize output 1. You should note that on the T1, the number of a particular input or output is written on the case of the PLC but for T2s and for some other more advanced PLCs this is not necessarily the case. To find out what the addresses of your inputs and outputs are you should refer to the documentation that came with your PLC. Also, in most ladder logic programming environments you have to specify the address of each of your inputs and outputs before it will even let you start programming (the T series can autoconfigure).

In ladder logic programming, you have registers instead of variables. There are four kinds of registers: Xs that are inputs; Ys that are outputs; Ds that are data that can form integers, hexadecimals, and real numbers; and finally Rs that are internal relays. Xs and Ys are pointers to the actual terminal strip connectors (what you use a screwdriver on to connect wires) on the PLC. If you energize an input, let's say 5, then X0005 will have an on status; also if you give Y0023 an on status, then relay 23 will flick on. Rs are just about the same as Xs and Ys except that they do not point to any hardware. They just hold an on or off value inside of the PLCs memory. Rs can be useful. Xs, Ys, and Rs can even hold data besides their on and off states on many PLCs, but personally I don't recommend it. For data like integers and hexadecimal numbers Ds are used as their addresses. Each of these instructions relates to a single bit of PLC memory that is specified by the instruction's address. Therefore, these instructions are also called *bit instructions*. A PLC may have several input modules and several output modules. Each of the modules takes one slot in the PLC frame. There are usually eight or more terminals in each module. During operation, a PLC processor may set (making logic HIGH) or reset (making logic LOW) a memory bit, based on the inputs and logical conditions of rungs in the program. In a ladder logic program, a relay logic instruction with the same address can be used as many times as the program needs.

Examine-if-Closed instruction. An Examine-if-Closed instruction is true when its memory bit is HIGH. The instruction is false when its memory bit is LOW. The logic relation between an Examine-if-Closed instruction and its memory status is similar to the relation between a normally open relay contact and the relay coil. The normally open contact is closed (true) when the relay coil is energized (HIGH). The normally open contact is open (false) when the coil is de-energized (LOW).

Examine-if-Open instruction. An Examine-if-Open instruction is true when its memory bit is LOW. The instruction is false when its memory bit is HIGH. The logic relation between an Examine-if-Open instruction and its memory status is similar to the relation between a normally closed relay contact and the relay coil. The normally closed contact is open (false) when the relay coil is energized (HIGH). The normally closed contact is closed (true) when the coil is de-energized (LOW).

Output Energize instruction. An Output Energize instruction is an instruction being controlled by other instructions. It can only be used at the position next to the right rail in a rung. When there is at least one path made by the instructions that are true from the left rail to the Output Energize instruction, this instruction is true. As long as a true path exists, the Output Energize instruction stays true. When there is not a true path, the Output Energize instruction is false.

Output Latch instruction. Like an Output Energize instruction, an Output Latch instruction needs a true path to get energized. After being energized, this instruction latches itself. Breaking the true path does not cause the energized Output Latch instruction to change to false. The memory bit assigned to an Output Latch instruction can be reset by an Output Unlatch instruction that has the same address as the Output Latch instruction.

Output Unlatch instruction. An Output Unlatch instruction is used for resetting an output instruction memory. It is usually used in conjunction with an Output Latch instruction. This instruction may only be located next to the right rail in a rung. When there is at least one path made by the instructions that are true from left rail to the Output Unlatch instruction, the instruction activates and resets the memory specified by its address.

Delay Timer. What a "delay timer" means is that after a specified amount of time after x0001 turns on, y0002 will turn on. You should note that because of the nature of ladder logic you cannot simply put a timer attached directly to the left-hand side without a relay conditional between it. Remember, everything is happening at the same time. PLCs are meant to run on their own for long periods of time, so you can't just tell it that 10 seconds after it is first plugged in it should activate something. You have to tell it to start timing after something in the outside world has occurred, like the energizing or de-energizing of an input. In the code --[01000 TON T012]-- there is the parameter 01000 that tells the timer to wait 1000*10ms or 10 seconds, and the parameter T012 tells the PLC which internal timer you want to use. Some of the more advanced PLCs have timers with different accuracy. Most measure time in 10 ms intervals but others measure time in single milliseconds. You should check the documentation on your PLC to see if any of its timers measure time in different units than the others. Also you should not use the same timer for more than one thing.

Conditional Statement. On rung three of the ladder we have a conditional statement. If the number stored in D0140 is equal to 1 then energize R001. If you look at the entire circuit you will note that there is nowhere else in it where D0140 is mentioned and you should know that all data registers are set to 0 at default. You may think that D0140 will never actually reach the value of 1 and that R001 will never be activated and that rung three and four are useless garbage code. It's true that during the normal operation of the PLC D0140 will never change from zero and the last two rungs before the end would be useless. However, this is where the computer link function comes in.

{END}. The final rung on the ladder, the -{END}-, is basically what it says. It is the end statement. It doesn't really do anything except to say, well, you're done programming. However, no program will work without an end statement and the PLC will ignore any code put in after an end statement. This shouldn't be a problem for small programs, just look at the screen and make sure the end is in there and at the bottom. If you happen to be making a very large and a very complicated relay circuit, your editor will likely force you to write it in separate blocks. Before attempting to write a very large program you should go to the very last programming block available to you and put the end statement there and nowhere else. The end statement can be used in debugging by ending the program early and disabling commands that fall after the end statement.

When a PLC is executing a ladder logic program, it examines the input modules first. It stores the status (HIGH or LOW) of the input devices, which are connected to the input modules, to memory bits assigned to the inputs. Then, the PLC scans the ladder logic program rung by rung from the top of the program to the bottom. During scanning, the PLC updates the statuses of the instructions in each rung. This will be discussed in the following paragraphs and later modules. After scanning the entire ladder logic program, the PLC copies the output instruction's memory statuses (HIGH or LOW) to the output modules to update the output terminals. The period from the beginning of the input terminal examination to the end of the output terminal updates is called a *scan cycle*. The time required for a scan cycle varies from program to program. It depends on the length of the program and the instructions used in the program. It may take 20 milliseconds or longer for each scan cycle. When a PLC is in the RUN mode, it executes the ladder logic program stored in its memory. The scan cycle is repeated continuously until the RUN mode is terminated.

When a PLC is scanning a rung, it looks for a path or paths made by the instructions, which are true, from the left rail to the instruction next to the right rail. If there is at least one true path, this rung is true and the instruction next to the right rail is set as true, or enabled. If there is not a true path, this rung is false and the instruction next to the right rail is set as false, or disabled. This instruction's memory is updated right after the PLC scans the rung.

Most of the functions will use PLC memory locations to get values, store values, and track function status. Most functions will normally become active when the input is true. But some functions, such as TOF (Timer OFF delay) timers, can remain active when the input is off. Other functions will only operate when the input goes from false to true; this is known as positive edge triggered. Consider a counter that only counts when the input goes from false to true, the length of time the input is true does not change the function behavior. A negative edge triggered function would be triggered when the input goes from true to false. Most functions are not edge triggered; unless stated, assume functions are not edge triggered.

11.3 Boolean Functions for Ladder Logic

Boolean algebra functions form the basis of ladder logic. The Boolean algebra function will obtain data words from bit memory, perform an operation, and store the results in a new location in bit memory. These functions are all oriented to word-level operations. The ability to perform Boolean operations allows logical operations on more than a single bit. The use of the Boolean

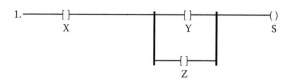

FIGURE 11.3
$S = X$ AND $(Y$ OR $Z)$.

functions is shown in Figure 11.3. Figure 11.3 realizes the function: $S = X$ AND $(Y$ OR $Z)$. For ladder logic

- "AND" is a two-place logical operation that results in a value of true if both of its operands are true, otherwise a value of false.
- "OR" is a logical operator that results in true whenever one or more of its operands are true. For example, in this context A OR B is true if A, B, or both A and B are true.

Typically, complex ladder logic is read left to right and top to bottom. As each of the lines (or rungs) is evaluated, the output coil of a rung may feed into the next stage of the ladder as an input. In a complex system there will be many rungs on a ladder, which are numbered in order of evaluation.

Figure 11.4 realizes $T = S$ AND X where S is equivalent to #1 above. This represents a slightly more complex system for rung 2. After the first line has been evaluated, the output coil (S) is fed into rung 2, which is then evaluated and the output coil T could be fed into an output device (buzzer, light, etc.) or into rung 3 on the ladder. (Note that the contact X on the second rung serves no useful purpose, as X is already defined in the AND function of S from the first rung.) This system allows very complex logic designs to be broken down and evaluated.

A more practical example, Figure 11.5, shows two key switches that security guards might use to activate an electric motor on a bank vault door. When the normally open contacts of both switches close, electricity is able to flow to the motor that opens the door. This is a logical AND.

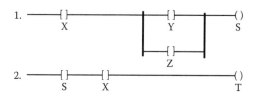

FIGURE 11.4
$T = S$ AND X where S is equivalent to #1 above.

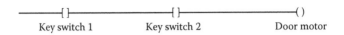

FIGURE 11.5
Two key switches to activate an electric motor.

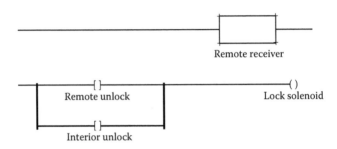

FIGURE 11.6
Two things that can trigger a car's power door locks.

Figure 11.6 shows the two things that can trigger a car's power door locks, which allow the driver or front passenger to simultaneously lock or unlock all the doors of an automobile or truck, by pressing a button or flipping a switch. The remote receiver is always powered. The lock solenoid, which converts energy into linear motion, gets power when either set of contacts is closed. This is a logical OR.

Often we have a start button to turn on a motor, and we want to turn it off with a stop button. Note the inverted logic of the stop function. The stop button is wired as a normally closed switch to the PLC input, which turns off when the stop button is pressed.

Figure 11.7 shows a latch configuration where we have a start button to turn on a motor, and we want to turn it off with a stop button. Note the inverted logic of the stop function. The stop button is wired as a normally closed switch to the PLC input, which turns off when the stop button is pressed.

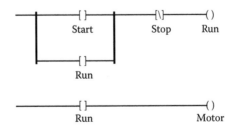

FIGURE 11.7
Start button to turn on a motor and stop button to turn it off.

11.4 Implement Jidoka Using Ladder Logic

The implementation of ladder logic is best understood by examining the historical perspective of their development. Figure 11.8 shows a simple electrical circuit to control a light with a single switch.

Let's introduce the symbols most often associated with programmable controllers. First, the switch is represented schematically, usually as just a pair of contacts, as shown in Figure 11.9. Most switches are meant to remain open, nonconducting, in their unactuated state. Figure 2 indicates such a switch. On the other hand, there may be reasons to have switches that are closed in their unactuated state, and open once they are actuated. Such switches are called normally closed (*NC*) and the symbol is shown in Figure 11.10.

In applications, both of these switches can be used to control a wide variety of loads: light bulbs, motors, buzzers, and so forth. The details of the load are unimportant to the logic of the control. Therefore, all loads can be represented by a generic circle, as shown in Figure 11.11.

Finally, the power supply is not explicitly represented in these diagrams but assumed to be common, in the background. A vertical line appears along the left-hand side of the page, representing the positive terminal of the power supply, while a similar line on the right-hand side represents the return, or negative side. This way, the electrical power appears to move left to right across the page. The simple circuit shown in Figure 11.8 can now be redrawn as shown in Figure 11.12.

In this manner, a circuit using switches and loads of arbitrary complexity can be drawn by inserting any elements between the two leads of the power supply. The diagram will simply "grow" down the page. This type of representation is called a ladder diagram and a single complete circuit is called a rung.

11.4.1 Logic Operations in Ladder Diagrams

Of course, if our objective was to control our loads with single switches, then there would be no need for a graphical language like ladder diagrams. The next step in complexity is the representation of logic operations in ladder diagrams.

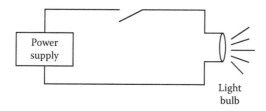

FIGURE 11.8
Simple single-switch control of a light bulb.

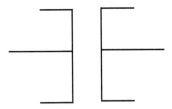

FIGURE 11.9
Schematic representation of a normally open (NO) switch.

FIGURE 11.10
Schematic representation of a normally closed (NC) switch.

FIGURE 11.11
Schematic representation of a load.

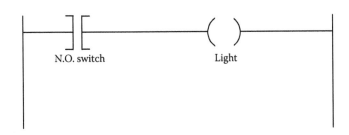

FIGURE 11.12
New schematic of simple light control circuit.

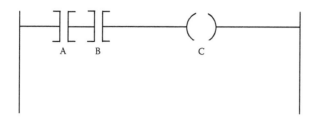

FIGURE 11.13
Ladder rung for $C = A$ AND B.

Suppose we have an application for which we want the load application only if two switches, switch A and switch B are activated. This is an instance of the logical AND operator. The load is activated if both A AND B are on. The rung in the ladder diagram would look like Figure 11.13. Similarly, if we wish the load C to be activated if either A OR B is activated, Figure 11.14 would apply.

Now let's consider a simple application. Back in the 1970s, automotive manufacturers implemented a feature called "seat belt interlock" in which the car could not be started if a person was sitting in the front seat and the seat belt was not fastened. Let's consider the ladder logic implementation of this system. First, we assume that there is always a driver seated behind the wheel. That would imply that the driver-side seat belt must always be fastened to allow the car to start. Let's represent the driver-side seat belt detection switch as D, a normally open switch that is closed when the seat belt is buckled. The circuit must be constructed such that D is closed before the engine ignition system, E, gets power. On the passenger side, it's a bit more complicated. We can only require that the passenger-side seat belt is buckled if there is a passenger seated. Let's consider two more switches, PS, a normally closed switch which opens when a passenger is seated, and PB, a normally open switch that detects whether the passenger seat belt is buckled. So, we allow power to the engine ignition system if the driver-side belt (D) is on AND there is no one in the passenger seat (PS closed) OR the driver side belt (D) is on AND the passenger-side belt (PB) is buckled. This combination of ANDs and ORs can be represented quite succinctly as shown in Figure 11.15.

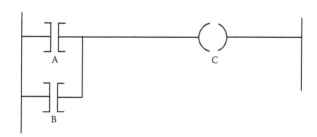

FIGURE 11.14
Ladder rung for $C = A$ OR B.

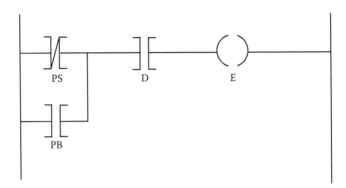

FIGURE 11.15
Ladder diagram implementation of seat belt interlock system.

11.4.2   Implementation of Ladder Diagrams

The ladder diagrams described in the previous section make up a formal graphical language to represent logical relationships between on–off inputs (e.g., switches) and on–off loads (e.g., lights, motors, buzzers, valves). While it is possible to implement the logical operations represented in ladder diagrams as electrical circuits, they are most often used as a programming language for PLCs. PLCs have many advantages when compared to hardwired implementation of ladder diagrams. Perhaps the most notable of which is that their programs can be easily modified and edited, whereas small changes in a hardwired implementation can be difficult and tedious to carry out.

PLCs are industrial-grade microprocessor systems consisting of a central processing unit and various input and output modules that can be interfaced directly to the machine's circuitry. Typically, there is also interface circuitry between the CPU and a traditional personal computer (PC). This interface is typically an RS 232 serial link, which allows for ladder logic programming in the PC environment. PLC programs are then compiled in the PC and downloaded through the serial line to the PLC. This arrangement also allows for off-line mass storage of PLC programs (on the PC).

11.5   Additional Functionality
for Manufacturing Autonomation

Additional functionality can be added to a ladder logic implementation by the PLC manufacturer as a special block. When the special block is powered, it executes code on predetermined arguments. These arguments may be displayed within the special block.

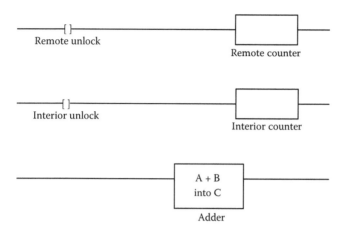

FIGURE 11.16
Ladder diagram implementation of a special lock.

In the example shown in Figure 11.16, the system will count the number of times that the interior and remote unlock buttons are pressed. This information will be stored in memory locations A and B. Memory location C will hold the total number of times that the door has been unlocked electronically.

PLCs have many types of special blocks. They include timers, arithmetic operators and comparisons, table lookups, text processing, proportional–integral–derivative (PID) controller control, and filtering functions. More powerful PLCs can operate on a group of internal memory locations and execute an operation on a range of addresses, for example, to simulate a physical sequential drum controller or a finite state machine. In some cases, users can define their own special blocks, which effectively are subroutines or macros. The large library of special blocks, along with high-speed execution, has allowed use of PLCs to implement very complex automation systems.

As shown by Figure 11.17, rung one is especially interesting: A TON and a TOF combination that lets output Y001 cycle on and off for 10 seconds at a time. While TON waits a given time before allowing an energized input to affect an output, TOF waits a given time before de-energizing an output after its input has been cut off. Let's analyze the rung:

- The -| |- conditional with X001 is there for good programming. It isn't actually necessary in this rung but if it's not there you have no way to stop the oscillating of Y001 during the PLC's operation. Now notice that we're not allowing current to flow if Y001 is energized yet the output of this rung is to energize Y001. Well, if Y001 is off then current is passed to TON. After TON has gone through its specified time, it will energize Y001. Now that Y001 is energized, current to the rung is cut off. Once the current has stopped TOF will keep Y001

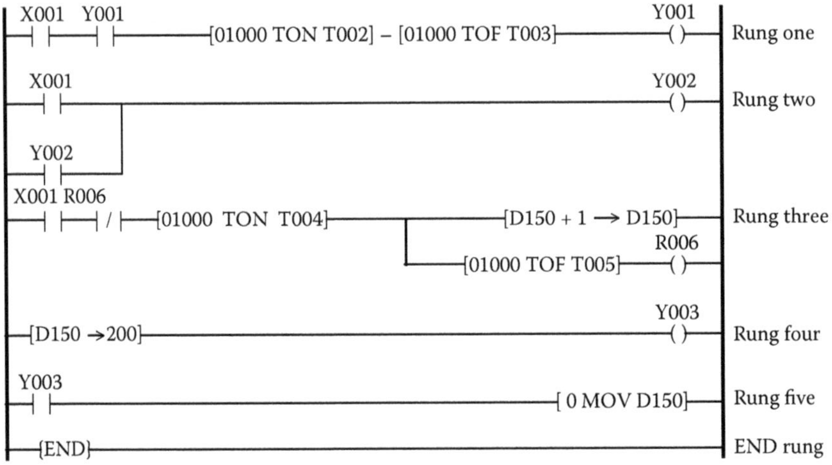

FIGURE 11.17
Implementation of a relay ladder logic.

powered for a specified amount of time before Y001 feels the affects and de-energizes. With Y001 de-energized TON is energized again and the cycle goes on. (You may want to reread that last sentence a few times.) The output Y001 stays on for a second, then off for a second continuously cycling.

- Rung two shows how to set up a push button. With a momentary push button, the input is only energized for a short amount of time, but often it is useful to keep the rung on long afterwards. What you should notice about rung two is that the output is connected to the left-hand side twice. If either of the two relay conditionals is on, then the output is on. Notice that one of the relay conditionals is the output itself. Thus if the output is powered for just one of the PLC's cycles (a very short time) then the output's own momentary on state will keep itself energized. You want to put in an internal relay conditional between the Y002 input and output or else you'll never be able to get it off! Well you could if you restart the PLC's program or with the computer link protocol.

- Rung three is also spread across two rung slots, but you could get it on one rung if it would fit (it really doesn't all fit on one rung in the editor). It's basically like rung one except that between the timers is a data statement that increments D150 by one. There are other ways to increment a data register, but this is what I used. Because it's between the timers, the data function will only get power and operate once during the period of the cycle. Since most of the timers only measure in milliseconds you can use a rung like this to measure time in hours or days if your PLC doesn't have any function that will do it for you.

- Rung four simply turns on an output when D150 is greater than a number, that is, a certain amount of time has gone by.
- Rung five sets D150 back to 0 when that output has been on momentarily. In a real application of something like this you'd probably want to use a TOF on rung four so that when D150 is no longer greater than 200, Y003 will wait a moment before deactivating, otherwise it will deactivate after one cycle of the PLC. The PLC probably goes through about a thousand cycles per second. A cycle is when the PLC updates the on, off, and value states of the relays and registers.

12

Pull System, One-Piece Flow, and Single Minute Exchange of Die (SMED)

12.1 Establish a Pull System

Does the office tell the factory floor not only what to build, but when to build it? If so, consider using a pull system. Pull is a basic concept of flow (Lean) manufacturing. It is at the center of how a site operates.

12.1.1 What Is a Push System?

A push system releases or schedules materials, items, or services into the process as projected customer orders are processed and materials become available. Push is generally anticipatory and often based on projected need (just-in-case).

Traditionally there is an office system that feeds the manufacturing floor with information not only about what to build but when to build it. It is a list of which product (customer) should be first, which product (customer) should be second, and so forth. This list is published daily. However, the list rarely takes into account the reality of the factory floor, for example, machine breakdowns, lost parts, and absenteeism. In a push system:

- Instructions come from a central planning area.
- Everyone's work is scheduled and triggered by required start dates.
- All work is assumed to take as long as allowed with no defective parts ever built.
- Since this never happens, time and inventory buffers are built to compensate.
- If a downstream operation has stopped, the upstream operation keeps on making product. Downstream is the work area you pass the product to when you have finished your part. Upstream is the work area from which you receive product on which to work.
- There is a lot of paperwork.
- If operators (or areas) run out of parts, they get chewed out, so they build extras and hide them in their toolbox.
- Some people do not want to hear this; however, it is common knowledge and should be dealt with. Flow exposes the rocks!

Example of pull:

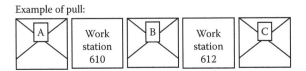

FIGURE 12.1
Illustration of a pull system.

12.1.2 What Is a Pull System?

A pull system controls the flow of work through an organization by only releasing materials, items, or services into the process as the customer demands them, that is, only when they are needed. The office still tells the shop floor what to build and in what order. However, the *when* to build comes from a downstream signal. This system creates only the parts requested from a downstream workstation. If there is no signal to build, the operator stops building.

Some characteristics of a pull system include:

- What to build and in what order still come from a central planning area.
- The receiving workstation does not accept any defective parts.
- Paperwork and computer transactions are greatly reduced.

A pull system is illustrated in Figure 12.1. If the operator at workstation 610 has completed his work and B is empty, the operator places the product in area B, then pulls a new product into workstation 610 to work on. If the operator at workstation 610 has completed his work and B still has product in it, then the operator leaves the completed product in workstation 610, then sees if station 612 needs help. Work stops at workstation 610 until the operator is able to place product into area B.

Lean = What the customer wants, when the customer wants it

12.2 Why Use Pull Systems and One-Piece Flow?

Appropriate use of a pull system is seen as key toward achieving true just-in-time (JIT) service provision. It is one of the three elements of just-in-time, along with takt time and continuous flow (using Kanban).

A pull system produces or processes an item only when the customer needs it and has requested it: use one, make one. The customer can be internal or external. It is an essential part of any build-to-order or provide-to-order

12.4   Single Minute Exchange of Die (SMED)

Single minute exchange of die (SMED) is one of the many Lean production methods for reducing waste in a manufacturing process. It provides a rapid and efficient way of converting a manufacturing process from running the current product to running the next product. It is also often referred to as quick changeover (QCO). Performing faster changeovers is important in manufacturing, or any process, because they make low cost flexible operations possible.

The phrase "single minute" does not mean that all changeovers and startup should take only one minute, but that they should take less than ten minutes (in other words, "single digit minute"). Closely associated is a yet more challenging concept of one-touch exchange of die (OTED), which says changeovers can and should take less than 100 seconds.

The SMED concept arose in the late 1950s and early 1960s, when Shigeo Shingo, chief engineer of Toyota, was contemplating Toyota's inability to construct vehicles in maximally efficient economic lots. The economic lot size is calculated from the ratio of actual production time and the changeover time, which is the time taken to stop production of a product and start production of the same or another product. If changeover takes a long time, then the lost production due to changeover drives up the cost of the actual production itself. This can be seen from Table 12.1 where the changeover and processing time per unit are held constant while the lot size is changed. The operation time is the unit processing time with the overhead of the changeover included. The ratio is the percentage increase in effective operating time caused by the changeover.

Toyota's additional problem was that land costs in Japan are very high and therefore it was very expensive to store economic lots of its vehicles. The result was that its costs were higher than other producers because it had to produce vehicles in uneconomic lots.

The economic lot size (or EOQ) is a well-known, and hugely debated, manufacturing concept. Historically, the overhead costs of retooling a process were minimized by maximizing the number of items that the process should construct before changing to another model. This makes the changeover

TABLE 12.1

Changeover Time

Changeover Time	Lot Size	Process Time Per Item	Operation Time	Ratio
8 hr	100	1 min	5.8 min	580%
8 hr	1,000	1 min	1.48 min	48%
8 hr	10,000	1 min	1.048 min	5%

overhead per manufactured unit low. According to some sources, optimum lot size occurs when the interest costs of storing the lot size of items equals the value lost when the production line is shut down. The difference, for Toyota, was that the economic lot size calculation included high overhead costs to pay for the land to store the vehicles. Shingo could do nothing about the interest rate, but he had total control of the factory processes. If the changeover costs could be reduced, then the economic lot size could be reduced, directly reducing expenses. Indeed the whole debate over EOQ becomes restructured if still relevant. It should also be noted that large lot sizes require higher stock levels to be kept in the rest of the process and these, more hidden costs, are also reduced by the smaller lot sizes made possible by SMED.

Over a period of several years, Toyota reworked factory fixtures and vehicle components to maximize their common parts, minimize and standardize assembly tools and steps, and utilize common tooling. This common parts or tooling reduced changeover time. Wherever the tooling could not be common, steps were taken to make the tooling quick to change.

EXAMPLE 12.1

The most difficult tooling to change was the dies on the large transfer-stamping machines that produce car vehicle bodies. The dies must be changed for each model. They weigh many tons and must be assembled in the stamping machines with tolerances of less than a millimeter.

When engineers examined the changeover, they discovered that the established procedure was to stop the line, let down the dies by an overhead crane, position the dies in the machine by human eyesight, and then adjust their position with crowbars while making individual test stampings. The process took from twelve hours to three days.

The first improvement was to place precision measurement devices on the transfer stamping machines and record the necessary measurements for each model's die. Installing the die against these measurements rather than by eye immediately cut the changeover to a mere hour and a half.

Further observations led to further improvements:

- Scheduling the die changes in a standard sequence as a new model moved through the factory.
- Dedicating tools to the die-change process so that all needed tools were nearby.
- Scheduling use of the overhead cranes, so that the new die would be waiting as the old die was removed.

Using SMED processes, Toyota engineers cut the changeover time to less than ten minutes per die, and thereby reduced the economic lot size below one vehicle. The success of this program contributed directly to JIT manufacturing that is part of the Toyota Production System. SMED makes load balancing much more achievable by reducing economic lot size and thus stock levels.

12.5 How to Implement Single Minute Exchange of Die (SMED)

Shigeo Shingo recognizes eight techniques that should be considered in implementing SMED.

1. Separate internal from external setup operations.
2. Convert internal to external setup.
3. Standardize function, not shape.
4. Use functional clamps or eliminate fasteners altogether.
5. Use intermediate jigs.
6. Adopt parallel operations (see Figure 12.2).
7. Eliminate adjustments.
8. Mechanization

SMED improvement should pass through the following four conceptual stages:

1. Ensure that external setup actions are performed while the machine is still running.
2. Separate external and internal setup actions.
3. Ensure that the parts all function and implement efficient ways of transporting the die and other parts.
4. Convert internal setup actions to external.
5. Improve all setup actions.

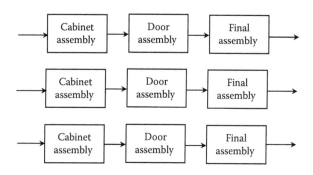

FIGURE 12.2
Adopting parallel operations to a refrigerator assembly.

12.5.1 Formal Method

There are seven basic steps to reducing changeover using the SMED system:

1. Observe the current methodology (A).
2. Separate the internal and external activities (B). Internal activities are those that can only be performed when the process is stopped, while external activities can be done while the last batch is being produced or once the next batch has started. For example, get the required tools for the job before the machine stops.
3. Convert (where possible) internal activities into external ones (C) (preheating of tools is a good example of this).
4. Streamline the remaining internal activities, by simplifying them (D). Focus on fixings: Shingo rightly observed that it's only the last turn of a bolt that tightens it; the rest is just movement.
5. Streamline the external activities, so that they are of a similar scale to the internal ones (D).
6. Document the new procedure and actions that are yet to be completed.
7. Do it all again: For each iteration of the above process, a 45-percent improvement in setup times should be expected, so it may take several iterations to cross the ten-minute line.

12.5.2 Key Elements to Observe

Look for and record all necessary data:

- Shortages, mistakes, inadequate verification of equipment causing delays, which can be avoided by check tables, especially visual ones, and setup on an intermediary jig
- Inadequate or incomplete repairs to equipment causing rework and delays
- Optimization for least work as opposed to least delay

TABLE 12.2

Operation and Proportion of Time

Operation	Proportion of Time
Preparation, after-process adjustment, and checking of raw materials, blades, dies, jigs, gauges, etc.	30%
Mounting and removing blades, etc.	5%
Centering, dimensioning, and setting of conditions	15%
Trial runs and adjustments	50%

- Unheated molds that require several wasted "tests" before they will be at the temperature to work
- Using slow precise adjustment equipment for the large coarse part of adjustment
- Lack of visual lines or benchmarks for part placement on the equipment
- Forcing a changeover between different raw materials when a continuous feed, or near equivalent, is possible
- Lack of functional standardization, that is, standardization of only the parts necessary for setup, for example, all bolts use same size spanner, die grip points are in the same place on all dies
- Too much operator movement around the equipment during setup
- More attachment points than actually required for the forces to be constrained
- Attachment points that take more than one turn to fasten
- Any adjustments after initial setup
- Any use of experts during setup
- Any adjustments of assisting tools such as guides or switches

12.5.3   Effects of Implementation

Shigeo Shingo, who created the SMED approach, claims that in his data from between 1975 and 1985 average setup times he has dealt with have reduced to 2.5 percent of the time originally required, a 97-percent improvement.

However, the power of SMED is that it has a lot of other effects, which come from systematically looking at operations. These include:

- Stockless production, which drives capital turnover rates
- Reduction in footprint of processes with reduced inventory freeing floor space
- Productivity increases or reduced production time
  - Increased machine work rates from reduced setup times even if number of changeovers increases
  - Elimination of setup errors and elimination of trial runs reduces defect rates
  - Improved quality from fully regulated operating conditions in advance
  - Increased safety from simpler setups
  - Simplified housekeeping from fewer tools and better organization
  - Lower expense of setups

- • Operator preferred since easier to achieve
- • Lower skill requirements since changes are now designed into the process rather than a matter of skilled judgment
- Elimination of unusable stock from model changeovers and demand estimate errors
- Goods are not lost through deterioration
- Ability to mix production gives flexibility and further inventory reductions as well as opening the door to revolutionized production methods (large orders ≠ large production lot sizes)
- New attitudes on controllability of work process among staff

12.6   Case Study: Flow Management of an Integrated Circuit Packaging Work-in-Process System

Integrated circuit (IC) manufacturing can generally be divided into five major parts: design, fabrication, chip probe testing, packaging, and final testing. IC packaging transforms wafers into chips. Since the product cycle time is short and the product arrival cannot be planned, IC packaging industry managers heavily rely on shop-floor information to timely handle dynamic shop-floor environments and respond to customer requests. Therefore, improving the competitiveness of a company mostly depends on flexibly incorporating shop-floor information into the production flow and handling manufacturing data quickly according to the production flow. A standard IC packaging production flow consists of the following operations:

Wafer grinding
Wafer sawing
Die bonding
Epoxy cure
Wire bonding
Post-bonding inspection (PBI)
Molding
Marking
Dejunking/trimming
Solder plating
Forming/singulation
Open/short testing
Lead scan

Final inspection

Packaging

Outgoing quality control

Typically recorded on a run card is the manufacturing information of the production flow for each lot is manually noted by floor personnel and later summarized by an assigned department. This card is filled out as the production batch arrives on the floor, recorded by the floor people based on the production facts at every operation, and finished when the production batch reaches completion. Although easy to process, the run card creates several problems:

- The exceptions on the floor cannot be effectively handled since such information is difficult to obtain timely.
- The best resolution time is missed due to the slow rate of accumulating time-critical information.
- The data is invalid due to human error.

To overcome the above problems and ensure good customer service, an IC packaging work-in-process (WIP) system is required. The WIP system should allow the user to perform the following tasks:

- Flexibly define shop-floor information in the production flow before production
- Timely handle the lot activities of customer orders according to predefined production flow during production
- Easily modify production flow information to allow fulfilling of changing product requirements and floor situations.

A WIP system is the core of the shop-floor information system, which bridges the planning and control levels. The shop-floor information system takes orders from the planning level, then coordinates, controls, and monitors equipment in the control level. A shop-floor information system includes the following eight functions: order review and release, detailed assignment, data collection and monitoring, feedback and corrective action, order disposition, scheduling, dispatching, and statistical process control.

The shop-floor information system can also be broadly defined as a manufacturing execution system (MES), which bridges manufacturing resource planning and the shop-floor supervisory system. In addition to integrating individuals, materials, machines, time, and cost via a relational database, this system also provides operation tracking, feeds information back to the planning system, and produces managerial reports for decision support.

12.6.1   Bill of Materials (BOM)

Narrowly defined, shop-floor information system uses a bill of materials (BOM) and routes in the planning level, and is combined with a shop-floor reporting system to monitor production flow. Thus, its application is limited by the lack of a flexible production flow data scheme capable of assisting the user to define what shop-floor information to collect and how to handle exceptions timely.

12.6.2   Data Mining

Data mining is a powerful technology used in the manufacturing industries to discover useful information. In these industries, a computer-integrated manufacturing (CIM) system has been used widely. Data-mining technology can be integrated with a CIM system to analyze the data of the real situation of the manufacturing process. Integrated data-mining technology is being integrated with statistical process control in CIM systems. The process yield rate can be improved through the use of automatic and optimum parameter manufacturing.

12.6.3   Computer-Integrated Manufacturing (CIM)

In the digital era, the population that accesses digital information has increased dramatically because of the widely accepted environment. Advanced and common information technology have become the necessary tools in the operation of enterprises and our daily life. In addition, information technology is also applied to lots of domains such as academia and industry. In academia, it aids students, teachers, and researchers to discover knowledge and solve problems on campus. CIM systems are widely used in modern manufacturing companies to improve the product quality and yield rate.

In the knowledge economy era, data-mining technology is a powerful tool for enterprises to identify useful information. In the manufacturing industry, the data-mining system and CIM system could be integrated to analyze the data of the manufacturing process. The production yield rate could be improved, and process cycle times, manufacturing costs, and inventory could be reduced. Therefore, an effective integration of information technology and algorithm, together with an appropriate decision-making model, could help the insufficiency in the analysis and prediction capability of the current management information system.

12.6.4   Semiconductor Manufacture Processes

In the modern semiconductor industry, to provide customers with high efficiency and high quality services became a great challenge to semiconductor manufacturers. For example, the derived product of an IC can meet the current requirements to be lightweight, thin, short, and compact. Along with the improvement in process equipment, most advanced production technology in the industry can perform a half line width of 0.13 micron. Owing to this superiority, which leads to the continuous increase of unit production capability of ICs, the unit prices get reduced rapidly. Hence, enterprise can reduce its materials and manufacturing cost effectively via new technology and continuous improvement on process capability and yield rate. In addition, the requirement of cost is no longer confined to the cost of equipment.

In addition, component services, running cost, and leading time are also very important. Therefore, semiconductor manufacturers need a powerful manufacturing system to aid the manufacture. The CIM system was designed to fit the requirements of manufacturers and could tighten production schedules, significantly increase productivity, reduce the required manpower, and lower expenditures. Production and manufacturing will test works that are all supported by the integrated computer system. It is possible to produce high-quality products with specs and reliability that meet customer's requirement. Therefore, CIM could improve the product quality. Furthermore, the WIP would be reduced inventory by using CIM system. It is possible to control waiting and setup time effectively and further reduce the preprocessing time in production and manufacturing.

12.6.5   Data Mining

The definition of data mining is to uncover useful information from a large amount of data. The purpose of data mining is to extract interesting knowledge from a database, data warehouse, or some other large information storage unit. It is an important step in the knowledge discovery in database processes. From a technical viewpoint, it combines the method of gathering and cataloging information then proceeds to generate rulelike knowledge from a large amount of data.

Data mining is a powerful new technology with great potential to help companies focus on the most important information in their data warehouse. Data mining scours databases based on hidden patterns, finding predictive information those experts may miss because it lies outside their expectations.

A particular data-mining algorithm is usually an instantiation of the model preference search components. The more common model functions in the current data mining process include the following:

- Association rules—Describes association relationships among different attributes.
- Classification—Classifies a data item into one of several predefined categories.

- Clustering—Maps a data item into a cluster, where clusters are natural groupings of data items based on similarity metrics or probability density models.
- Regression—Maps a data item to a real-valued prediction variable.
- Summarization—Provides a compact description for a subset of data.
- Dependency modeling—Describes significant dependencies among variables.
- Sequence analysis—Models sequential patterns, like time-series analysis. The goal is to model the states of the process generating the sequence or to extract and report deviations and trends over time.

12.6.6 Ant Colony Optimization

The ant colony optimization (ACO) algorithm is one of the data mining methods used in the industry. ACO is a probabilistic technique that is used to find near-optimal solutions to combinatorial optimization problems. It was originally inspired by the natural system that ants use to develop a short path between their colony and a food source. As an ant moves, it leaves behind a pheromone trail. The movement of any ants following that first ant depends on the detection of the pheromones on that trail. Not only will the following ants detect and follow that trail, they will also seek newer and better paths based on the amounts of pheromones detected. This pheromone trail can be presented as a numeric value. Therefore, using these numbers, we should be able to calculate and set parameters for pheromone values.

A problem is NP-hard if solving it in polynomial time would make it possible to solve all problems in class NP in polynomial time. Ant colony algorithm has been successfully applied to several NP-hard problems. Just as its name implies, the ACO algorithm originates from the study of the behavior of a natural ant colony. There are three ideas from the natural ant colony that has been transferred to the artificial ant colony:

1. The preference for paths with a high pheromone level
2. The higher rate of growth in the amount of pheromones on shorter paths
3. The information exchanged among ants

12.6.7 Run-to-Run Control

Semiconductor manufacturing involves highly complicated processes and high machine costs. To maintain stable product quality, enhance manufacturing efficiency, improve product quality and reliability, the most effective way was to control and monitor the manufacturing process automatically. Therefore, there were some methods of process control applied to manufacturing processes. In the manufacturing process, process variances might occur due to some unknown reason; therefore, the output of the process might become unstable. Statistical process control (SPC) is the most common method used in the manufacturing process. SPC is a statistical-based approach that monitors long-term process performance. Engineers would define the upper or lower spec limit of manufacturing parameters or define the upper or lower control limit of manufacturing parameters. Then, manufacturing engineers could trace these parameters to assure the good yield rate and define the boundary of manufacturing parameters.

13

Lean Manufacturing Business Scorecards

13.1   What Are Lean Manufacturing Business Scorecards?

Lean manufacturing business scorecards (LMBSs) are interactive computer-based structures and subsystems intended to help decision makers use communication technologies, data, documents, knowledge, and analytical models to identify and solve problems. The new generation of LMBSs offers the potential for significantly improving operational and strategic performance for organizations of various sizes and types.

For example, industrial motor-driven systems consume more than 70 percent of global manufacturing electricity annually and offer one of the largest opportunities for energy savings. System optimization techniques through the application of existing, commercially available technologies and accepted engineering practices typically achieve energy savings of 20 percent or more for these systems across all industrial sectors. The optimization opportunities for steam systems are at least equal or greater.

Despite the potential benefits, energy savings from these industrial systems have remained largely unrealized by U.S. industry. This chapter presents a Lean manufacturer practitioner's argument that unless energy efficiency is institutionalized, it will be viewed by corporate managers as something different than the effective and efficient use of labor and material resources. If this institutionalization does not occur, the potential benefits will never be achieved or sustained. The LMBS can be used to identify and sustain the potential benefits.

The same factors that make it so challenging to achieve and sustain energy efficiency in industrial motor-driven and steam systems (complexity, frequent changes in production processes and staff) apply to the production processes that they support. Yet production processes typically operate within a narrow band of acceptable performance. These processes are frequently incorporated into quality and environmental management systems (ISO, Six Sigma, others), which require measurement, documentation, and continuous improvement. During the 1990s, most large organizations engaged in enterprise data warehousing projects. The scope of these efforts ranged from combining multiple legacy systems to developing user interface tools for analysis and reporting. The data warehouse is the underlying structure that is used to generate a variety of reports and analyses. In the past, business intelligence amounted to a set of weekly or monthly reports that tended to be unconnected.

This chapter will present the experience of Marion Appliance Company in integrating energy efficiency into the strategic business plan via the Marion production system and other operating practices. Starting with an internal champion at Marion who is also a Six Sigma Black Belt, I will describe how building on several existing management structures (Marion strategic business plan, Marion production system, including Marion total productive maintenance), and implementing actions using common operating practices (total value management, Marion best practice systems, and Marion Six Sigma) creates an environment that encourages a more energy-efficient operation. First, the chapter will explore how energy efficiency has been incorporated into the business plan and operational procedures. Next, examples will be given of how sharing best practices, procedures, and successful projects via these systems are resulting in measurable improvements. Then finally, a specific model for integrating a compressed air system into Six Sigma will be described.

13.2 How Have Lean Manufacturing Business Scorecards Been Evolved?

Two salient features of the new generation of LMBSs are integration and visualization. Typically, this information flow is presented to the manager via a graphics display called a dashboard. An LMBS dashboard serves the same function as a car's dashboard. Specifically, it reports key organizational performance data and options on a near real time and integrated basis. Some LMBS industry pundits claim that dashboards are simply "eye candy" for executive managers. This perspective suggests that these systems are merely a new fad being promoted by consultants and vendors. Although these claims may have some merit, dashboard-based business intelligence systems do provide managers with access to powerful analytical systems and tools in a user-friendly environment. Furthermore, they help support organization-wide analysis and integrated decision making.

The first LMBSs actually went into organizations around 1985. They were called manufacturing information systems at the time. And they had limited success because they were executive systems—the chairman of Merck would have it on his desk—but then that was it. What we're seeing today are management dashboards, which have been pushed down through the organization, providing relevant information to a particular manager. At Southwest Airlines, they call them cockpits and they're specialized so that the guy in charge of putting peanuts on airplanes gets a different view than the guy who's in charge of purchasing jet fuel. But they all see where planes are flying.

The optimization of industrial motor-driven and steam systems offers an outstanding and typically overlooked opportunity to improve the bottom line. Experience in the United States, Canada, United Kingdom, and China has demonstrated that energy efficiency improvements of 20 to 25 percent are typical and often can be as high as 50 percent over existing conditions. These are operational cost reductions that frequently result in additional benefits such as increased reliability, tighter process control, lower maintenance costs, and reduced waste. System optimization can also yield increased capacity to support production without a significant capital outlay through more effective utilization of existing equipment.

If system optimization is so beneficial, why isn't U.S. industry already doing it? There are several factors that contribute to a widespread failure to recognize this opportunity, not just in the United States but globally. These include the complexity of these systems and the institutional structures within which they operate. Industrial systems are ubiquitous in the manufacturing environment, but their applications are highly varied. They are *supporting* systems, so facility engineers are typically responsible for their operation, but production practices on the plant floor (over which the facility engineer has little influence) can have a significant impact on their operational efficiency. System optimization cannot be achieved through simplistic, one-size-fits-all approaches. Both industrial markets and policy makers tend to focus on equipment components (motors and drives, compressors, pumps, boilers), which can be seen, touched, and rated, rather than focus on systems, which require engineering and measurement. The presence of energy-efficient components, while important, provides no assurance that an industrial system will be energy efficient. In fact, the misapplication of energy-efficient equipment (such as variable speed drives) in these systems is common. The disappointing results from these misapplications can provide a serious disincentive for any subsequent effort toward system optimization.

System optimization requires taking a step back to determine what work (process temperature maintained, production task performed, etc.) needs to be performed. Only when these objectives have been identified can analysis be conducted to determine how best to achieve them in the most energy-efficient and cost-effective manner.

The skills required to optimize systems are readily transferable to any individual with existing knowledge of basic engineering principles and industrial operations. Training and educational programs in the United States and the United Kingdom have successfully transferred system optimization skills since the early 1990s. Plant engineers with an awareness of the benefits of system optimization still face significant barriers to achieving it.

First, existing systems were typically not designed with operational efficiency in mind. As a result, basic design factors such as pipe size may be too expensive to retrofit and may require a work-around approach to do the best optimization project possible.

Second, once the importance of optimizing a system and identifying system optimization projects is understood, plant engineering and operations staff frequently experience difficulty in achieving management support. The reasons for this are many, but central among them are two: (1) a management focus on production as the core activity, not energy efficiency; and (2) lack of management understanding of operational costs and equipment life-cycle cost, which is further exacerbated by the existence of a budgetary disconnect in industrial facility management between capital projects (including equipment purchases) and operating expenses.

Third, as a further complication, experience has shown that most optimized systems lose their initial efficiency gains over time due to personnel and production changes. Since system optimization knowledge typically resides with an individual who has received training, detailed operating instructions are not integrated with quality control and production management systems.

Since production is the core function of most industrial facilities, it follows that the most sophisticated management strategies would be applied to these highly complex processes. Successful production processes are consistent, adaptable, resource efficient, and continually improving—the very qualities that would support industrial system optimization. Because production processes have the attention of upper management, the budgetary disconnect between capitals and operating budgets is less evident. Unfortunately, efficient use of energy is typically not addressed in these management systems in the same way as other resources such as labor and materials. I think the answer lies in fully integrating energy efficiency into these existing management systems.

13.3 Lean Manufacturing Business Scorecard for Planning and Policy Deployment

Typically, LMBSs can be categorized into two major types: model driven and data driven. Model-driven systems tend to utilize analytical constructs such as forecasting, optimization algorithms, simulations, decision trees, and rules engines. Data-driven systems deal with data warehouses, databases, and online analytical processing (OLAP) technology.

A data warehouse is a database that is constructed to support the decision-making process across an organization. There may be several databases or data marts that make up the data warehouse. Managers utilize OLAP to help process and evaluate large-scale data warehouses and data marts increasingly. In five years, 100 million people will be using information-visualization tools on a near daily basis. And products that have visualization as one of their top three features will earn $1 billion per year.

Today, there is an ongoing requirement for more precise decision making because of increased global competition. Generally speaking, decision making should be based on an evaluation of current trends, historical performance metrics, and forecast planning. New and improved LMBSs continue to emerge to help meet these ongoing requirements. Within three years, users will begin demanding near-real-time analysis relating to their business in the same fashion as they monitor stock quotes online today. Monthly and even daily reports won't be good enough. Business intelligence will be more focused on vertical industries and feature more predictive modeling instead of ad hoc queries.

Lasting change is evolutionary not revolutionary. Mandating change will yield only short-term results. To obtain sustainability, the change must become part of the culture and a way of doing business. Most energy efficiency actions are in the form of projects, programs, and initiatives. While these actions are important to jump start energy efficiency, without the plan to institutionalize these actions, the savings will be short lived. For example, installing a new energy management system will help reduce energy as long as the system, controls, and equipment are maintained. Weekend shutdown programs will only work as long as someone is assigned to shut down equipment and follow-up verification is done. Initiatives, such as reducing compressed air system pressure can only be supported as long as the associated equipment specifications are modified to reflect the lower pressure requirements.

At Marion Appliance Company, energy efficiency was institutionalized by making it part of the way business was conducted through business plans, policy deployment, and the Marion production system (MPS). Actions are deployed and problems solved in the same manner production and business issues are resolved, using methods like Six Sigma, best practice replication (BPR), total value management (TVM) teams, and single point lessons (SPLs). This has allowed energy efficiency to become not only an additional task to do, but also a part of the way Marion does business. The following will reflect on the processes and how energy is imbedded in the process. Although some of the terminology used in this chapter is unique to Marion Appliance Company, the organizational concepts and implementation mechanisms are generic and can be applied anywhere.

As with every company, this process starts with a vision, a business plan, and an implementation process. In this case the vision is "great products, strong business, better world." To fulfill this goal, there are eight strategic priorities and the business plans support these priorities. From the overall vision, philosophically energy conservation fits: Strong business by reducing costs through lower usage and a better world by reducing emissions through lower energy supply requirements. The question is how does this fit into real-world practice. Although it can be argued that energy can fit into almost all of the eight strategic priorities, the two that it has been institutionalized into is sustainability, which focuses on a better environment and a better society, and Lean manufacturing. The core of Lean manufacturing, as in energy conservation, is the elimination of waste. For the remainder of this chapter, the concentration will be on how energy is ingrained in Lean manufacturing. Although the examples will be specific to vehicle operations, the principles, processes, and procedures are the same for all of manufacturing and form the core of the manufacturing business plan.

13.3.1   Business Plan Process

Before one can understand how energy efficiency is built into the system, a basic understanding of the process is necessary. As previously stated, the Marion business plan is a set of "roll-up" actions and objectives from each area that feeds into meeting the business model and the company's targets and objectives. The North American business model is established along with the North American score card (the Y priorities). The plans are established using a Y = f(X) model. In this case it would be manufacturing establishing its critical Xs that support the North American Ys. The business plan defines the Xs and the targets, while the manufacturing scorecard establishes and tracks the metrics. A master schedule is developed as a timeline for action implementations and to track progress to the key milestones to ensure that the organization is on track to meet its goals. This same process is repeated in the divisions, into the plants, and down to the production-team level where each action taken rolls up as the input to the next level's output and alignment is created from the production floor all the way to the CEO. In vehicle operations this process is accomplished through the use of manufacturing councils and policy deployment.

13.3.2   Manufacturing Councils

In manufacturing, the plans and objectives are broken into six major categories: safety, quality, delivery, cost, morale, and environmental (SQDCME). For each of these elements, a manufacturing council acts as the governing body, developing the business plan, monitoring progress, and assisting in implementation. This is a cross-functional group that is championed by one of the directors of manufacturing and the division subject matter expert (SME) and includes plant managers, Lean manufacturing managers, and support staff. They take the objectives and develop actions and metrics that support achieving the objectives. Because of the composition of the councils, these actions become an integral part of operations and can be deployed without the normal resistance that occurs if objectives are dictated. After the "catch ball" sessions throughout the levels of the organization, the objectives are cascaded throughout the organization to the work groups.

13.3.3   Policy Deployment

Once the business plan process is complete, we use policy deployment to cascade the objectives down to the work groups. The SQDCME objectives are cascaded to the plant via the actions and the plant scorecard. Each plant will break these down by area, zone, and then by work groups. At this level, the work group team leaders and supervisors develop and track their inputs and actions. The team objectives and actions are tracked by the team leader on the team board and scorecard, and are reviewed with the team during their team meeting. Plant management reviews team boards on a daily basis to see what actions the teams need help with.

Energy is embedded in two areas of the scorecard. The first area is under cost. It is one of the items in FUFA (fuel, utilities, and fixed assets), which is part of the plant's budget and is tracked as a part of the total cost. The second area is environmental, where energy usage is a stand-alone item on the scorecard. On the vehicle operations (VO) division scorecard and on each plant scorecard, the 2005 objective is a 20-percent reduction in energy based on the year 2000 baseline. The reduction target is then divided by the area and given to the work groups. The production work groups develop specific actions in their area, such as tracking and shutting off unused lights, fans, and other equipment. The skilled trades teams develop weekend and daily shutdown plans and manage the leak tag program. These actions are what the plant can control to reduce energy; the objectives are set at every level of the organization. This process has led to VO plants meeting or exceeding the 20-percent reduction objective in 2005.

13.3.4   Marion Production System (MPS)

The Marion production system is the method of doing business in manufacturing. It is a standardized process with operational elements and procedures that govern the way manufacturing and business is run. It is also the method that must be used to transform Marion from mass production to Lean manufacturing. It is about cultural change and, most importantly, engaging the work force. There are eleven elements in MPS:

- Safety and health assessment review process (SHARP)
- Environmental
- Leadership
- Work groups
- Training
- In-station process control (ISPC)
- Marion total productive maintenance (MPM)
- Manufacturing engineering (ME)
- Synchronous material flow (SMF)
- Industrial material flow (IMF)
- Quality operating system (QOS)

These elements are the DNA of the system; the plants go through an annual validation process where they are scored on their adherence to and their advancement in the process. Energy is interwoven into the process and is a part of the validation and the scoring. It is directly and indirectly included in a number of the elements but the two elements that have the strongest impact on energy are the Marion total productive maintenance and manufacturing engineering elements.

13.3.5 Total Productive Maintenance (TPM)

TPM is the element that governs the maintenance of the equipment and processes in the plant. An effective maintenance program is the key to continued energy efficiency and optimum equipment performance. Some examples of the effects of poor maintenance are:

- Dirty filters increasing fan load
- Poor relamping program decreases watts/lumens thus increasing electric usage
- Air leaks increase electric usage
- Faulty steam traps and steam leaks increase fuel consumption
- Fouled inlet filters on air compressors increase electricity consumption
- High inlet temperatures increase air dryer loads
- Fouled condensers increase chiller and air conditioner loading
- Inoperable/poorly calibrated controls reduce the effectiveness of an energy management system

TPM is the method used to ensure that the preventative maintenance system is in place and that the maintenance is getting done. We use a total equipment maintenance (TEM) computerized maintenance system to track and schedule maintenance. Plants are measured not only to their adherence in the use of the system but also to the completion of the inspections and work orders.

There is also a specific requirement in several of the TPM sections and procedures for adherence to a leak tag program that includes compressed air and steam. So a system has been put in place that maintains the equipment efficiency and reduces energy.

13.3.6 Manufacturing Engineering (ME)

The manufacturing engineering element includes energy in the reliability and maintainability (R&M) section as part of the total cost requirement, in the project review process where every new project has to include the effect on energy usage, and as a stand-alone section, energy management efficiency program.

13.3.7 Energy Management Efficiency Program

This section of the MPS ME element evaluates compliance to the energy management procedures and reviews specific actions to increase energy efficiency for the facility. The following are some of the questions asked:

1. Does the facility report monthly energy and utility usage data on a quarterly basis to Marion Land in a hard-copy report or electronically using the energy utilities and waste metrics (EUWM) Web site?
2. Does the facility report total value management (TVM) energy data on a weekly basis to Marion Land electronically using the TVM energy program Excel datasheet?
3. Has the plant reduced nonproduction electrical demand between shifts to 50 percent of normal production levels and to 25 percent during weekend and holiday periods?
4. Has the plant reduced nonproduction compressed air demand to 25 percent of normal production levels?
5. Has the plant reduced paint humidity in solvent-based paint booths to 50 percent during heating season?

Having this constantly tracked and monitored not only gives visibility but accountability to reducing energy. Again, the key is to include energy efficiency in the everyday process and the result is continuous improvement in energy efficiency.

13.4 Lean Manufacturing Business Scorecards for Problem Solving and Continuous Improvement

LMBSs provide necessary links and end-user interface for managers to access and receive selective information such as competitor behavior, industry trends, and current decision options. To increase organizational acceptance and use, these new systems feature distributed decision making, which helps leverage organizational visibility. Specific attention is being given to the user interface as highlighted by the following list of standard end-user features:

- Filter, sort, and analyze data
- Formulate ad hoc, predefined reports and templates
- Provide drag and drop capabilities
- Produce drillable charts and graphs
- Support multiple languages
- Generate alternative scenarios

13.4.1 Dashboards

There are a number of approaches for linking decision making to organizational performance. For example, in the manufacturing industry, decisions may focus on resource allocation optimization and waste reduction, as supported by the Lean manufacturing methodology. From a decision-maker's perspective, the new LMBS visualization tools such as dashboards and scorecards provide a useful way to view data and information. Outcomes displayed include single metrics, graphical trend analysis, capacity gauges, geographical maps, percentage share, stoplights, and variance comparisons. A dashboard-type user-interface design allows presentation of complex relationships and performance metrics in a format that is easily understandable and digestible by time-pressured managers. More specifically, such interface designs significantly shorten the learning curve and thus increase the likelihood of effective utilization. Figure 13.1 presents an example of a dashboard design.

13.4.2 Scorecards

A scorecard is a custom user interface that helps optimize an organization's performance by linking inputs and outputs both internally and externally. (The balanced scorecard is the specific methodology associated with the Kaplan and Norton model.) To be effective, the scorecard must link into the organization's vision. Over the next few years the differences between dashboards and scorecards will become increasingly blurred as these interface structures become fully integrated. Figure 13.2 illustrates the current adoption of LMBSs throughout the organization.

The dashboard integrates the data warehouses and analytical models directly into the decision making process. This is a continuous process based on ongoing environmental scanning and feedback from current performance metrics, for example, inventory turns. Behind the graphical interface lie the

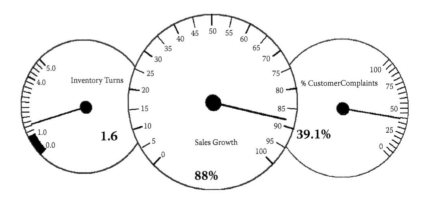

FIGURE 13.1
Example of a dashboard.

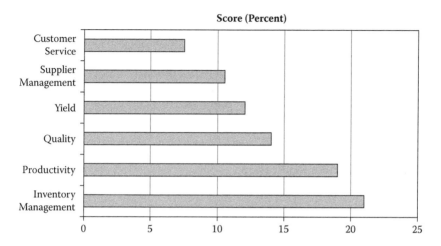

FIGURE 13.2
LMBS adoptions by management area.

supportive analytical systems such as statistical analysis for data validation, combined forecasting algorithms, and expert systems for decision options analysis and recommendations.

The first piece was including energy in the business plan, the second piece was to include it in the production plan (MPS), and the final piece of institutionalizing energy efficiency is to have it in the operational plan and treated like any other part of the business. As such, teams work on reducing energy by using the same tools and methods as they would with any other problem or objective.

13.4.3 Total Value Management (TVM)

The cross-functional TVM teams work on increasing the value of specific items and areas of cost. One of these teams, led by the energy department in Marion Land, is dedicated to reducing energy usage. It meets once a week in the energy "war room" to review progress to objectives and implementation of actions.

13.4.4 Single Point Lessons (SPLs)

The energy TVM has developed and deployed SPLs on reducing energy. SPLs were developed as part of MPS to get concise information and direction to the workforce, both hourly and salary. SPLs are papers, usually one page, that have a targeted and pointed message. It is a tool used to educate and instruct. The energy SPLs have centered on plant shutdown and there is currently one under development for air leaks.

13.4.5   Shutdown Monitoring

The energy TVM also monitors the effectiveness of the plant's weekend shutdowns. It gathers the data and publishes the percent of production demand that the plants were able to achieve. There is a monthly report that tracks the plant's progress to the energy objective and does a better–worse comparison to the previous year. This is used to populate the plant scorecard previously mentioned in the chapter. The key point here is that it is very important to measure and track the progress.

13.4.6   Best Practice Replication

Another process to facilitate energy reduction is best practice replication. A Web site has been established where plants, divisions, or communities submit their best practice ideas for replication. The Web site is divided into communities and the ideas have to be approved and verified by the "gatekeeper" of the community and by the subject matter experts. Once it is approved as a best practice, each plant must review the idea and implement it if possible. This accelerates the implementation process and aids in rapid deployment of these ideas. Energy has its own community, and energy best practices are found throughout the other communities. Currently there are over 150 best practices that relate to energy on the Web site. The main point is not the Web site but the rapid deployment of these ideas through manufacturing. This process prevents wasted steps from everybody trying to reinvent the wheel. We call it "stealing with pride" when we use another facility's idea.

13.4.7   Six Sigma

Six Sigma is a statistical problem-solving tool that is commonly used in the industry. Using the DMAIC (define, measure, analyze, improve, control) process, Black Belts and Green Belts take on projects and problems, and find solutions. For example, while doing my Black Belt certification, I used the DMAIC process to analyze a problem with a compressed air system.

13.4.7.1   Define

The first step was to define the problem. The supplier, input, process, output, customer (SIPOC) chart and the CT (critical To) tree and a fishbone diagram were used to identify all of the inputs into the process. From there, the $Y = f(x)$ chart was developed, which takes the identified inputs and links them to the outputs. The $Y = f(x)$ chart helps identify the inputs that can be controlled, like compressor controls, compressor mix, and air leaks as well as those that are less controllable such as environmental conditions, electrical energy, and plant demand. Once the controllable inputs have been established, the process flow map can be accurately drawn and a cause and effect (C&E) matrix developed. The C&E matrix ranks the inputs and the potential outputs based on the importance to the customer. The higher the rating, the more important the effect. In this case, the top items were downtime, system cost, and process capability. The final define step was to do an FMEA (failure modes and effects analysis). This takes the information from the C&E matrix and, based on history, establishes the likelihood of the occurrence. In this case, while the compressor failure was the most important item on the C&E matrix, it was the least likely to occur, which affected its ranking on the FMEA. On the other hand, compressor controls/compressor mix and air leaks became more important due to their impact on system cost and their likelihood of occurrence. With this knowledge, the scope of the project could be narrowed and the measure phase started.

13.4.7.2 Measure

The following was done during the measure phase:

- Gathered CFM/KWH data on CTs by building
- Compared and analyzed data (sanity check)
- Developed efficiency measurement
- Normalized data using averages
- Established baseline

13.4.7.3 Analyze

Once the data was gathered and then normalized, descriptive statistics could be used to establish the curve and sigma value for the compressor utilization and develop the following:

- Narrow scope of project to body and final
- Compare compressor operation patterns
- Low efficiency—too many compressors on line
- Analyze mix of compressors
- Centrifugals in body more efficient than rotary screw machines in final
- Evaluate losses in system
- Develop and implement corrective actions

13.4.7.4 Improve

Once the actions were developed from the analyze phase, the following was done in the improve phase:

- Determine right mix of compressors for load
- Run test
- Evaluate system operation

- Develop answers to the "yeah but" list
- Shut down redundant compressor
- Evaluate savings
- Move to control/monitor phase

13.4.7.5 Control

During the control phase the savings were validated and the plan was developed to ensure that the systems stayed tied together and that the body compressors would be used for base load production. As a result, this project achieved a system efficiency improvement by combining the compressed air systems in two buildings and operating the most efficient compressors based on the compressed air demand. This project yielded a savings of over $236,000 with little to no investment.

There are over 200 Black Belt projects either completed or underway in the system that relate to energy. The key message is to use the tools that are available to solve problems and apply them to energy reduction.

13.5 Building the Business Intelligence Strategy

Training at all levels is a key ingredient in the successful application of LMBSs. In many applications, training occurs at the last minute and is simply geared toward how to use the system for specific assignments. Intensive training before, during, and after system implementation helps create the cultural change needed to maximize acceptance throughout the organization. Training simulators represent one approach for both improving system utilization and increasing organizational buy in. Within two to three years, companies will ditch the traditional model of making business adjustments on a quarterly basis. Instead, they'll use business intelligence and performance management tools to make real-time shifts in strategy to respond to changes in the marketplace. Some current technical challenges facing this evolving industry are:

- Integrating optimization-based models with enterprise resource planning systems
- Developing an observation-oriented approach to data modeling that includes manual and automated processing
- Combining decision support, knowledge management, and artificial intelligence in a data warehousing framework
- Designing intelligent agents that can be used to support decision-making processes
- Formulating adaptive and cooperating systems that use evaluation and feedback to improve the decision-making process

Additionally, speech recognition represents a significant development for improving the human–computer interface. Specifically, a speech interface system would allow the manager to increase the decision-making flow volume as well as to explore a broader range of unstructured decision applications. In the next two years, business intelligence capabilities will become more democratized, with a far greater number of end users across the enterprise using the tools to get better visibility into the performance of their segment of the business. Think of it as executive dashboards for worker bees.

Highlighted next are some specific examples in which dashboards have been successfully applied to improve organizational performance.

- Product development management—Historically, measuring the performance of ongoing product/service development (PD) has been a hit-or-miss proposition. This inconsistency has often led to significant overruns and, in some cases, total failure. Estimating the product development cycle time is key to any effective assessment process. A typical PD dashboard system is designed to report results to date as well as to indicate the potential for continuing success or failure. Project compliance is one key dashboard PD metric. A gauge reports the fraction of new product launches that occurred on schedule and budget. Another standard dashboard metric is the fraction of products or services that have received a favorable trade journal review. Additionally, the dashboard should have the capability of identifying new product or service opportunities.

- Production financial management—Many production organizations have concluded that it is essential to have real-time updates of key performance metrics such as revenues and profits in order to remain competitive in today's marketplace. Traditionally, many organizations have relied on quarterly reports to support the decision making process, a practice which has often led to uneven performance. A production financial dashboard provides an integrated and real time overview of performance that can be directly correlated to the business model. Specific metrics include balance sheets, income statements, and competitor performance. Additionally, the dashboard can display alerts identifying negative trends that require immediate attention.

Each of these applications was developed based on a well-designed business intelligence strategy. Developing an effective business intelligence strategy is predicated on three key drivers: perceived value, organizational utilization, and a cost-effective solution. The development of an LMBS strategy should be tied to specific organizational performance goals and operational objectives. Examples of the latter include increasing customer retention and reducing turnover of key personnel. The proposed solution must be adaptable, scalable, and maintainable. Often a phased schedule in implementing the LMBS is best since it tends to minimize risk as well as increase organizational acceptance. Such an approach allows elements of the system to be checked out prior to full system deployment. Presented in the following list are the major steps involved in developing an effective LMBS strategy:

- Establish LMBS objectives (specifically, what do you want the system to do)
- Evaluate the current in-house support capability, including the present system's architecture
- Perform a gap analysis on existing data systems, including response time
- Identify alternative technical solutions
- Formulate an implementation timeline
- Conduct organizational "town hall" meetings to solicit ideas and to enhance the cultural climate for change
- Determine the need for outsourcing support

Outsourcing some or all of the implementation process can offer significant benefits to organizations with limited internal technical capabilities or an already strained information technology (IT) department. Outsourcing also brings the latest in technological development. A first step when considering outsourcing is to assess the organization's internal infrastructure. This assessment is essential since LMBS applications can become very expensive whether developed internally or outsourced. The initial investment for developing an LMBS ranges from $1 million to $20 million-plus, depending on organizational goals, current IS capabilities, and the projected number of users. The annual system operating expenses can often equal a significant portion of the initial investment.

- The use of LMBSs throughout most organizations is on the increase as a result of growing global competitive pressures. Improved user friendliness through the use of graphic interfaces is a primary characteristic of the new generation of LMBS applications. Specifically, managers require interactive interface systems, such as dashboards, that are easy to understand and use. Organizational integration represents another important characteristic of LMBSs.
- Current industry challenges include improving system integration and developing cooperative and adaptive systems that incorporate feedback and evaluation automatically into the decision making process. More specifically, real-time speech recognition represents a new technology for improving the human–computer interface that is essential for use by managers at all levels.

- Developing an LMBS strategy involves three key issues: perceived value, organizational utilization, and a cost-effective solution. The development of an LMBS strategy should be tied to specific organizational performance goals. With a carefully crafted plan, organizations can realize significant increases in productivity and insights into the marketplace.

- Ongoing management training is essential for insuring the continued effective use of the LMBS. Simulation is one training strategy that provides an effective and dynamic structure for introducing and supporting new LMBS applications. Many organizations should consider outsourcing for implementing their BI strategy. The initial investment for an LMBS can range from $1 million to $20 million-plus, depending on the specific operational requirements.

- Some potential implementation barriers include failure to establish viable performance metrics, failure to fund adequate postsystem training, and failure to obtain organizational buy in.

The key to lasting energy efficiency and continuous improvement is to institutionalize the process. Energy efficiency has to become a part of doing business and not just an additional task that has to be done. Energy has to be tracked and monitored with specific goals and objectives that need to be reached. At Marion, energy has been included in the business plan, with targets set and cascaded throughout the organization. It then is tracked monthly on the plant and division scorecards. The people on the plant floor own the inputs that drive the results. The user becomes empowered to take the actions to reduce energy. To ensure that it remains institutionalized, energy has been included in the operating procedures and the same problem-solving tools that are used to solve other manufacturing problems are used to reduce energy. This further ingrains energy efficiency in the system because it becomes a problem that a team can correct by using the tools already known and does not take any additional training or resources. *The key is not to depend on projects and programs for energy reduction but to rely on processes and business practices.* This requires a company to make energy efficiency part of the standard practice; when energy use starts to increase, the company needs a process that flags it as an off-standard condition just like any other production standard. Remember, lasting change is evolutionary not revolutionary. To achieve lasting energy efficiency, it has to be part of the culture and a way of doing business.

Bibliography

Albano, R. E., Freedman, D. J., and Hendryx, S. A. (1990). "Manufacturing Execution Equipment," *AT&T Tech Journal*, 69(4), 53–63.

Anderson, D. M. (1998). *Agile Product Development for Mass Customization: How to Develop Products for Mass Customization, Niche Markets, JIT, Build-to-Order and Flexible Manufacturing*. New York: McGraw-Hill.

Babson, S. (Ed.). (1995). *Lean Work: Empowerment and Exploitation in the Global Auto Industry*. Detroit, MI: Wayne State University Press.

Black, J. T. (1996). *The Design of the Factory with a Future*. New York: McGraw-Hill.

Bowen, H. K., and Spears, S. (1999). "Decoding the DNA of the Toyota Production System," *Harvard Business Review* (September–October), 96–106.

Boyer, M., and Moreaux, M. (1997). "Capacity Commitment Versus Flexibility," *Journal of Economics & Management Strategy*, 6, 347–376.

Brandon, J. A. (1996). *Cellular Manufacturing: Integrating Technology and Management*. New York: John Wiley.

Burbidge, J. L. (1989). *Production Flow Analysis for Planning Group Technology*. Oxford, UK: Oxford University Press.

Carlsson, B. (1989). "Flexibility and the Theory of the Firm," *International Journal of Industrial Organization*, 7, 179–203.

Cheng, T. C. E., and Podolsky, S. (1993). *Just-In-Time Manufacturing: An Introduction*. London: Chapman & Hall.

Cochran, D. (1999). "The Production System Design and Deployment Framework," SAE technical paper 1999-01-1644, SAE IAM-99 Conference.

Cusumano, M. A. (1994). "The Limits of 'Lean,'" *Sloan Management Review*, 35(4), 27–32.

Dertouzos, M. L., Lester, R. K., and Solow, R. M. (1989). *Made in America: Regaining the Productive Edge*. Cambridge, MA: MIT Press.

Dhavale, D. G. (1996). *Management Accounting Issues in Cellular Manufacturing Systems and Focused-Factory Systems*. Montvale, NJ: The IMA Foundation for Applied Research Inc.

Eaton, B. C., and Schmitt, N. (1994). "Flexible Manufacturing and Market Structure," *American Economic Review*, 84, 875–888.

Ericksen, M., Powers, E., & Ribeiro, F. (2007). "Trimming Organizational Costs … For the Long Term." *Financial Executive*, 23, 48–51.

Ericsson, A., and Erixon, G. (1999). *Controlling Design Variants: Modular Product Platforms*. Dearborn, MI: Society of Manufacturing Engineers.

Feigenbaum, A. V. (1991). *Total Quality Control* (3rd ed.). New York: McGraw-Hill, Inc.

Flinchbaugh, J., and Carlino, A. (2006). *The Hitchhiker's Guide to Lean*. Dearborn, MI: Society of Manufacturing Engineers.

Fujimoto, T. (1999). *The Evolution of a Manufacturing System at Toyota*. New York: Oxford University Press.

Gallagher, C. C., and Knight, W. A. (1986). *Group Technology Production Methods in Manufacture*. New York: John Wiley.

Gerwin, D. (1993). "Manufacturing Flexibility: A Strategic Perspective," *Management Science*, 33, 395–410.

Goldratt, E. M. (1990). *The Theory of Constraints*. New York: North River Press.

Goldratt, E. M. (1992). *The Goal* (2nd rev. ed.). Great Barrington, MA: North River Press.

Gross, J. M., and McInnis, K. R. (2003). *Kanban Made Simple*. New York: American Management Association.

Hedden, R. P. (1987). *Cost Engineering in Printed Circuit Board Manufacturing* (Cost Engineering Series, Vol. 11). Boca Raton, FL: CRC Press.

Hopp, W. J., and Spearman, M. L (2001). *Factory Physics: Foundations of Manufacturing Management* (2nd ed.). New York: McGraw-Hill Higher Education.

Humphreys, J. (2005). "Developing the Big Picture," *MIT Sloan Management Review*, 47, 95–96.

Hyer, N. L., and Wemmerlov, U. (2002). *Reorganizing the Factory: Competing through Cellular Manufacturing*. Portland, OR: Productivity Press.

Irani, S. A. (Ed.). (1999). *Handbook of Cellular Manufacturing Systems*. New York: John Wiley.

Jiang, J. C., and Chen, K. H. (2007). "Development of a Collaborative Manufacturing, Planning, and Scheduling System: Integrating Lean and Agile Manufacturing for the Supply Chain," *International Journal of Management*, 24(2), 331–345.

Jones, R. A., and Ostroy, J. M. (1984). "Flexibility and Uncertainty," *Review of Economics Studies*, 64, 13–32.

Kamrani, A. K., Parsaei, H. R., and Liles, D. H. (Eds.). (1995). *Planning, Design and Analysis of Cellular Manufacturing Systems*. New York: Elsevier Science B.V.

Kaplan, R. S., and Norton, D. P. (1996). "Using the Balanced Scorecard as a Strategic Management System," *Harvard Business Review*, 74(1), 75.

Kennedy, F., Owens-Jackson, L., Burney, L., and Schoon, M. (2007). "How Do Your Measurements Stack Up to Lean?" *Strategic Finance*, 88(11), 32–41.

Lee, H., Padmanabhan, P., and Whang, S. (1997). "Information Distortion in a Supply Chain: The Bullwhip Effect," *Management Science*, 43, 546–558.

Lee, R. N. (Ed.). (1992). *Making Manufacturing Cells Work*. Dearborn, MI: Society of Manufacturing Engineers.

Levy, D. L. (1997). "Lean Production in an International Supply Chain." *Sloan Management Review*, 38, 94–102.

Liker, J. K. (2004). *The Toyota Way*. New York: McGraw-Hill.

Maynard, M. (2003). *The End of Detroit: How the Big Three Lost Their Grip on the American Car Market*. New York: Doubleday.

Monden, Y. (1998). *The Toyota Management System*. Portland, OR: Productivity Press.

Moodie, C., Uzsoy, R., and Yih, Y. (Eds.). (1995). *Manufacturing Cells: A Systems Engineering View*. London: Taylor & Francis.

Nash, M., and Poling, S. R. (2007). "Strategic Management of Lean," Quality, 46(4), 46–49.

Ohno, T. (1988). *Toyota Production System: Beyond Large-Scale Production*. Portland, OR: Productivity Press.

Pine, B. J., II. (1993). *Mass Customization: The New Frontier in Business Competition*. Boston: Harvard Business School Press.

Pine, B. J., II, and Boynton, A. (1993). "Making Mass Customization Work," *Harvard Business Review*, 71(5), 108–119.

Rother, M., Shook, J., Womack, J., and Jones, D. (1999). *Learning to See: Value Stream Mapping to Add Value and Eliminate Muda*. Brookline, MA: Lean Enterprise Institute.

Rubrich, L., and Watson, M. (1998). *Implementing World Class Manufacturing*. Fort Wayne, IN: WCM Associates.

Sarker, B. R., and Harris, R. D. (1988). "The Effect of Imbalance in a Just-In-Time Production System: A Simulation Study," *International Journal of Production Research*, 26(1), 1–18.

Sayer, N. J., and Williams, B. (2007). *Lean for Dummies*. Indianapolis, IN: John Wiley & Sons.

Schonberger, R. J. (1983). "Integration of Cellular Manufacturing and Just-In-Time Production," *Industrial Engineering*, 15(11), 66–71.

Sekine, K. (1990). *One-Piece Flow: Cell Design for Transforming the Production Process*. Cambridge, MA: Productivity Press.

Shingo, S. (1989). *A Study of the Toyota Production System from an Industrial Engineering Viewpoint*. Portland, OR: Productivity Press.

Singh, N., and Rajamani, D. (1996). *Cellular Manufacturing Systems*. New York: Chapman & Hall.

Sobek, D. K., II, Liker, J. K., and Ward, A. C. (1998). "Another Look at How Toyota Integrates Product Development," *Harvard Business Review*, 76(4), 36–49.

Spear, S., and Bowen, H. K. (1999). "Decoding the DNA of the Toyota Production System," *Harvard Business Review*, 77(5), 97–106.

Standard, C., and Davis, D. (1999). *Running Today's Factory: A Proven Strategy for Lean Manufacturing*. Dearborn, MI: SME.

Sterman, J. (2000). *Business Dynamics: Systems Thinking and Modeling for a Complex World*. New York: Irwin/McGraw-Hill.

Steudel, H. J., and Desruelle, P. (1992). *Manufacturing in the Nineties: How to Become a Mean, Lean World-Class Competitor*. New York: Van Nostrand Reinhold.

Suh, N., Cochran, D., and Lima, P. (1998). "Manufacturing System Design," *Annals of 48th General Assembly of CIRP*, 47(2), 627–639.

Toyota Motor Corporation. (1995). *The Toyota Production System*. Toyota City, Japan: Toyota Motor Corporation, Operations Management Consulting Division and International Public Affairs Division.

Vives, X. (1993). "Information, Flexibility and Competition," *Journal of the Japanese and International Economies*, 7, 219–239.

Wang, J. X. (2002). *What Every Engineer Should Know About Decision Making under Uncertainty*. Boca Raton, FL: CRC Press.

Wang, J. X. (2005). *Engineering Robust Design with Six Sigma*. Prentice Hall, NJ: Upper Saddle River.

Wang, J. X. (2008). *What Every Engineer Should Know About Business Communication*. CRC Press, FL: Boca Raton.

Wang, J. X., and Roush, M. L. (2000). *What Every Engineer Should Know About Risk Engineering and Management*. Boca Raton, FL: CRC Press.

Wenger, E. (1998). *Communities of practice, Learning, meaning, and identity*. Cambridge (U.K.): Cambridge University Press.

Womack, J. P., and Daniel, T. (1994). "From Lean Production to the Lean Enterprise," *Harvard Business Review*, 72(2), 93–103.

Womack, J. P., Jones, D. T., and Roos, D. (1991). *The Machine That Changed the World: The Story of Lean Production*. New York: First Harper Perennial.

Index